CONSEQUENCES OF CLIMATIC CHANGE

Edited by

Catherine Delano Smith and Martin Parry

Department of Geography
University of Nottingham

1981

TABLE OF CONTENTS

PART I

PART II

INTRODUCTION

The papers in this collection are the product of a conference of the Historical Geography Research Group of the Institute of British Geographers, which was convened at the University of Nottingham in July 1980. All but two of the papers (Barber and Delano Smith) were given at the conference or are based on comments from the floor. The substantial papers in Part I are based on the study of primary historical sources and reflect the authors' preoccupation with the identification of specific effects of climatic change on specified forms of human endeavour. They are not intended as a comprehensive indication of the 'state of the art' nor are they all refined and water-tight reports of completed research. They are, though, intended to indicate the paths which have been followed and the progress which has been made in evaluating the impact of changes in climate on early economies and societies. The papers in Part II represent short statements of research in progress or more personal comments on wider issues. Their value, we believe, lies in challenging too facile acceptance of unproven (and unprovable) relationships; in the setting up of a second opinion; or as an underlining of what cannot be overstated - the need for balanced consideration of the role of all factors, human as well as physical, in changes in economies and societies in the past. In as much as environmental change is hardly a standard component of orthodox enquiry in social and economic history, it should not be surprising that some of the arguments presented here are not orthodox while some might even be described as provocative. At the same time, it should be emphasised that the authors have not been concerned to provide here the balanced explanation of historical events just alluded to or to attempt to weigh the relative impacts of a number of different casual factors. Obedient to their conference brief, they have focussed on what each would readily admit is only one of a range of elements in an explanation. The aim has been to seek to comprehend more fully the role of climatic change before proceeding to evaluate it in the context of the more orthodox factors of historical change.

It was decided to publish these papers as an attempt to link two extremities of the research spectrum in the context of climatic change - the archaeological and the climatological. Historical geographers have a stance somewhere between, in touch with each. Archaeologists, faced with the hard evidence of a past event, seek an explanation. In certain areas, notably marginal ones, or in certain contexts, notably those of major changes or discontinuities in settlement, the temptation to ascribe almost all to climatic change as a convenient or compelling *deus ex machina* is not always resisted. In part this reflects uncertainty as to the detailed mechanics of climate/agricultural relationships or of the role, even relevance, of the numerous non-climatic factors in an economic or social situation. Help is hardly at hand, though, at the other end of the spectrum, where climatologists - armed with modern, highly technical data - discuss in polished but bewildering terms atmospheric relationships that would be more familiar to a physicist than to a historian. It is hoped, therefore, that the following papers will serve to illustrate various paths of enquiry that seek to narrow the area of uncertainty surrounding the relationship between climate and history.

It would be very wrong however, for the reader to infer from the emphasis on climatic change in this collection that the authors believe climatic change to have been the most important single factor affecting the phenomena they have studied. The truth is that we are not yet able to compare effectively the relative impact of social, economic and environmental forces for change. Nor should the title of this collection *Consequences of Climatic Change* be taken to imply an assumption on our part either that climates *have* changed significantly or that their consequences *exist* and our task has been merely to evaluate these consequences. Even these are uncertainties. The papers in this collection do, however, draw attention to some areas of agreement in the evidence.

ACKNOWLEDGEMENTS

We are greatly indebted to a number of people who helped to make the production of these papers possible. Some of the illustrations were drawn in the Department of Geography, University of Birmingham, by Carl Burness, Jean Dowling and Tim Grogan. Others were drawn in the Department of Geography, University of Nottingham, by Christopher Lewis, who also did the cover work. The typing was done by Tracey Allen, Pat Davidson and Rosemary Hudson. Also in the same department, John Winn carried the burden of all the photographic and reprographic work. On the financial side, help has been received from the Historical Geography Research Group and from the Department of Geography, University of Nottingham. To all these, we extend our warm thanks.

C.D.S.
M.L.P.

February 1981

EVALUATING THE IMPACT OF CLIMATIC CHANGE

M. L. Parry
Department of Geography
University of Birmingham

There has been a tendency to assume that the space-time coincidence of climatic change and economic change indicates a causal connection. This premise is false, and its associated research strategy is weak. One way of establishing more precisely the contingent relationships between climate and economy in the past is to adopt a 'retrodictive' strategy. This is demonstrated and exemplified by analysis of the connection between climatic cooling and the abandonment of upland cultivation in early modern Britain.

The purpose of this paper is to outline a research strategy for evaluating the climatic change on economic history. There is a need for a new strategy because our enquiries have met with little success so far. There may be two explanations for this. Firstly, economic historians have traditionally displayed a curious ambivalance to climatic change. They have been confused by the quite different biological and agronomic effects of short-term variability and long-term changes of climate. It is not surprising that their comparisons of short-term variations (the kind perceived by contemporaries, remarked upon in their diaries and for which evidence survives) and long-term agricultural change appear to show little connection. They have also failed to appreciate that climatic variations have quite different effects in different places. A long-term fluctuation of 1°C might have little effect on the growing conditions of summer in a continental temperature regime but would markedly alter conditions in a maritime area such as in northern and western Europe. Such a change would, for example, return central England to average temperatures prevailing in the so-called "Little Ice Age" and it would be misleading to describe this scale of fluctuation as "slight, almost intangible" (Le Roy Ladurie, 1972).

In addition, there has been a tendency for historians to cling to superseded chronologies. For example, that proposed by Brooks (1949) is still quoted as evidence that the early medieval period was "one of Europe's cold cycles" (Postan, 1975). Since this period was one of settlement expansion it has been concluded that village desertions had little to do with climate (Beresford, 1971). In fact, more recent chronologies see a phase of more frequently cool and wet summers being coincident with the time of most widespread desertion, A.D. 1370-1500 (Lamb, 1977; Parry, 1978); this is not to say that there is any causal connection between the two events, but it at least suggests a line of worthwhile enquiry.

A second explanation for lack of success in evaluating the role of climate change in economic history is the failure of climatologists to grasp the peculiar constraints imposed by historical problems on the deductive path of enquiry. Most historical explanations are probabilistic in that historians can hope to account for only some of the "contingent conditions" for occurrences (Nagel, 1961). Their statements have an uneliminable subjective component which does not lend itself readily to analysis in the strictly deductive form in which the climatologist is trained. An outcome of this is that some climatic historians have despairingly resorted to an investigation of the synchroneity of climatic and economic events, and to assume that events which occur synchronously and in the same place are causally related. Yet in several studies which have adopted this line of reasoning there is little evidence that climatic change was a *contingently necessary condition* of coincident economic change.

An example may illustrate this point. It has been suggested that the decline of Mycenaean culture was the result of intense and prolonged drought around 1000 B.C. Two points are made in support of this argument:

1) that winter drought has occurred during the period of meteorological observations in the area (i.e. for a few weeks over December-January 1954/55) - and thus could have occurred in the past.
2) that the pattern of weather events around 1000 B.C. in eastern Europe and the eastern Mediterranean fit broadly with a drought in the Peloponnese. In other words, a drought could have occurred (Bryson, Lamb and Donley, 1974).

But the premise of a connection between reduced rainfall, famine, and out-migration is not examined. In essence all that is being argued is that drought and out-migration were coincident, the assumption being that if they were coincident then they were connected. There has been no attempt to model the processes which supposedly connected the decrease in rainfall, declining crop yields and the decision to move. The argument incorporates all the weaknesses of inductivism: it lacks an unbroken logic which establishes the relationships between the various propositions and the propositions have not been tested empirically or theoretically.

One means by which we may establish the contingent relationships between climate and economy in the past is to adopt a deductive approach, one which makes retrodictions about the past and tests them against historical actuality.

A RETRODICTIVE STRATEGY

Predicted reconstructions of the past which can subsequently be tested against the historical record, cannot readily be made for social and economic phenomena because of the changeable nature of societies and economies. But it is possible to analyse the present physiological relationship between climate and crops and, by allowing for differences in the type of crop and in crop husbandry, to adopt the relationship as an analogy of processes in the past (Fig. 1). Any retrodictions based upon the analogue model of process links will be a statement about a functional relationship, not simply a statement of space - time

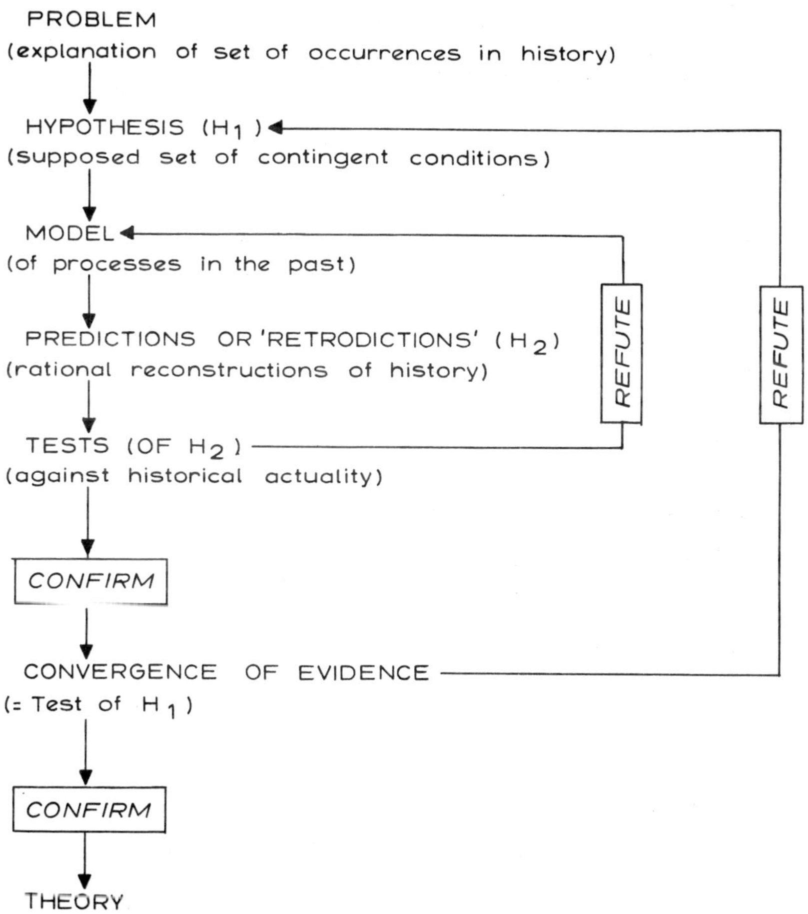

Figure 1 - A 'retrodictive' strategy in historical research.

coincidence. If the retrodiction is not confirmed then either the model is incorrect, or the hypothesis cannot be supported. If the retrodiction is confirmed, and if there is converging evidence from independent sources, we may say the hypothesis itself can be confirmed (Fig. 1). It is thus possible to employ the power of the predictive technique of deductivism in climate-history studies. The remainder of this paper will explain and illustrate how this may be done.

The illustration will be drawn from a particular problem - the need to explain extensive and permanent abandonment of high-level farmland in northern Europe in the late medieval and early modern periods (Fig. 2). The explanation might be sought in:

1) changes in farming objectives (e.g. in the minimum tolerance levels of crop yields; objectives adjusted to meet changing demand);
2) in changes in farming systems (particularly technological changes which gave to other areas, previously unfarmed, a comparative advantage over high-level farmland);
3) errors of farming judgement. Some high-level reclamation, particularly during the exuberant expansion of agriculture over 1780-1815, was simply an over-optimistic (and expensive) mistake which it took no more than one or two cool summers to reveal;
4) climatic change leading to changes in the physiological viability of farming and to changes in productivity.

In the final analysis each of these explanations must be weighed against each other, but in this illustration the discussion will focus on one alone - the premise that permanent land abandonment is somehow connected with long-term climatic change. Following the broad path of the retrodictive approach, the task is, firstly, to model the processes linking climate and farming decisions in the past, and, secondly, to test predictions derived from the model.

MODELLING CLIMATE-FARMING PROCESSES

There are four sets of processes that need to be analysed (Fig. 2):

1) the relationship between time, scale and causality in climate-farming interactions;
2) analogous crop-climate processes in the present,
3) the relationship between weather variability and farming decisions;
4) changes in agro-climate.

Time, scale and the nature of climate-farming interactions.

The premise that permanent land abandonment was connected with long-term climatic change is based upon four assumptions. Firstly, a distinction is drawn between long-term changes which might be expected to have long-term effects, and short-term changes which probably connect with short-

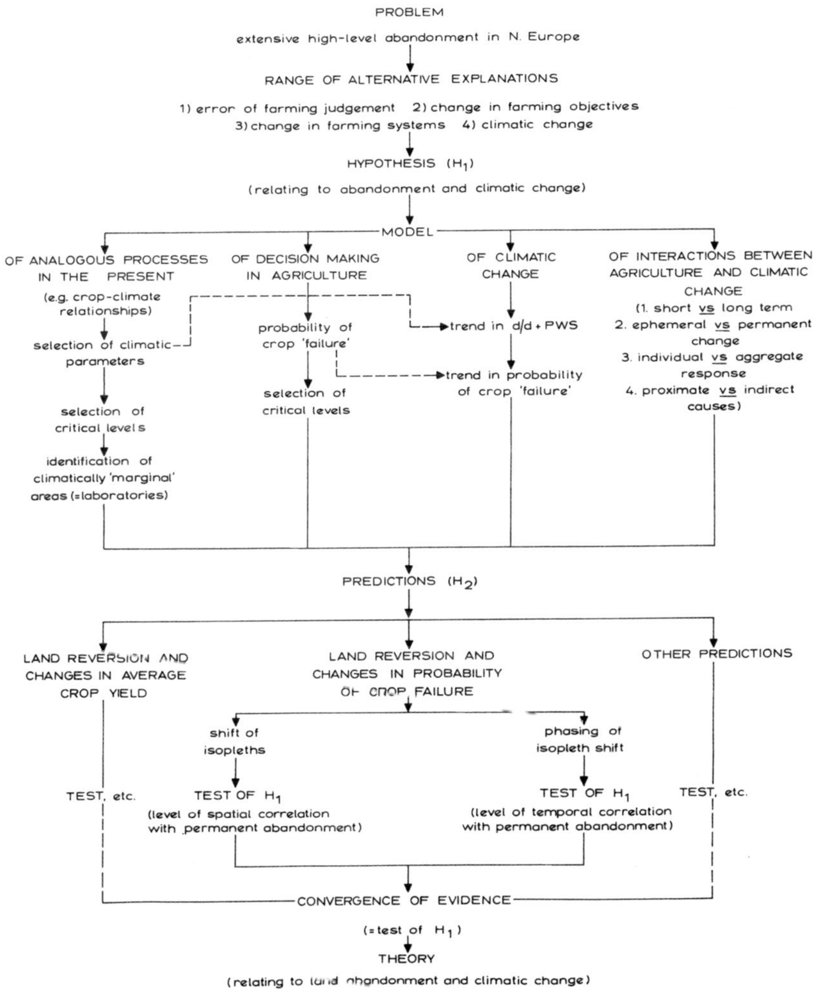

Figure 2 - The retrodictive strategy: climatic change and land abandonment. Note: d/d = summer warmth in day-degrees above 4.4°C; PWS = end of season potential water surplus.

term effects unless they trigger a response that has been preconditioned by long-term factors. For example, a short run of harsh years cannot be a sufficient condition for abandonment "in the long run", i.e. cannot explain the failure to re-settle an "abandoned" region after the return of more benign weather. There must be other explanations that account for the longevity of the response.

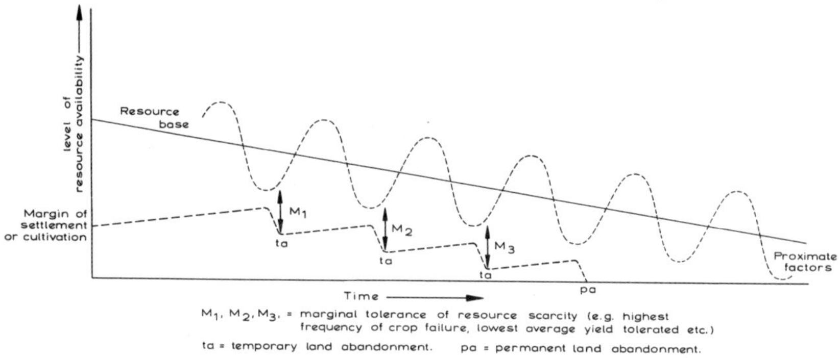

Figure 3 - Proximate and indirect factors behind land abandonment and settlement desertion.

Secondly, a distinction has been drawn between ephemeral (or temporary) agricultural response and permanent agricultural response; and between single events which are the result of individual decisions (e.g. the abandonment of single farms) and aggregate events which are the result of similar decisions by many men (e.g. the abandonment of many farms over a large area).

Finally we have distinguished between hypotheses of climatic change as proximate or precipitating causes of agricultural change and those which are underlying or indirect causes of agricultural change.

The branches of these dichotomies (short-term/long-term; ephemeral /permanent; individual/aggregate; proximate/indirect) are interlocked. For example, if the concern is to explain the long-term abandonment of farmland which was the aggregate product of the decisions of many farmers, then the concern will be with long-term or step-wise climatic changes (since some farmers may survive short runs of harsh years but will go down with the rest over the long-term), and with climatic changes as underlying or indirect "causes" of farm abandonment. Similarly, if the concern is to explain ephemeral events such as the temporary abandonment of a few farms, then the interest will focus on short-term climatic variability as a proximate or precipitating factor.

Long-term climate change can thus be seen as inducing long-term changes in the resource base for agriculture. The level of resource availability of marginal agriculture varies with shifts in the resource base and the variation of proximate factors (prices, technology, epidemics, wars, short-run weather impacts, etc.) that play on this

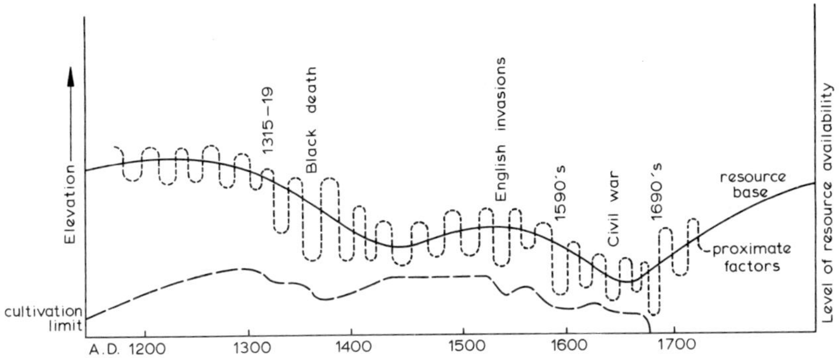

Figure 4 - Proximate and indirect factors behind farmland abandonment in upland Scotland: a schematic chronology

shifting stage (Fig. 3). Proximate factors may encourage ephemeral retreat when the margin of tolerance of resource scarcity is temporarily exceeded, but extensive and permanent retreat is likely to occur only when long-term changes in the resource base, sometimes as a result of climatic change, have reduced the stage upon which the proximate factors play. The argument is that, without this secular decline, the proximate factors would have induced a less permanent and less widespread result.

Figure 4 shows, in a highly schematised form, the interaction of proximate factors and the shifting stage of summer warmth and summer wetness in Scotland over A.D. 1200-1700. The proximate factors are the stuff of orthodox history - political upheaval, the Black Death, the decline of the monasteries, or the runs of disastrous summers in the 1590s and 1690s - and these were the factors perceived by contemporaries whose comments survive in the records. Such factors are no doubt the major explanation for land abandonment, but the degree to which agriculture was responsive to such factors depended upon its marginality - and its marginality could be seriously affected by decline in the resource base. We may thus venture that the abandonment of high-level farmland was 'triggered' by proximate factors acting on a marginality enhanced by climate changes in the fourteenth and seventeenth centuries.

Analogous crop-climate processes

With some knowledge of time and scale in the interaction between climate and past agricultural change, we now need to establish precisely those parameters of climate to which the farming system was most sensitive. This can be achieved by adopting present-day crop-climate relationships as anologues of those in the past.

Let us assume that high-level cultivation was based largely on the cropping of oats, a fair assumption for northern Britain. Both early and modern varieties of oats are particularly sensitive to exposure, summer warmth and summer wetness. These may be measured by average

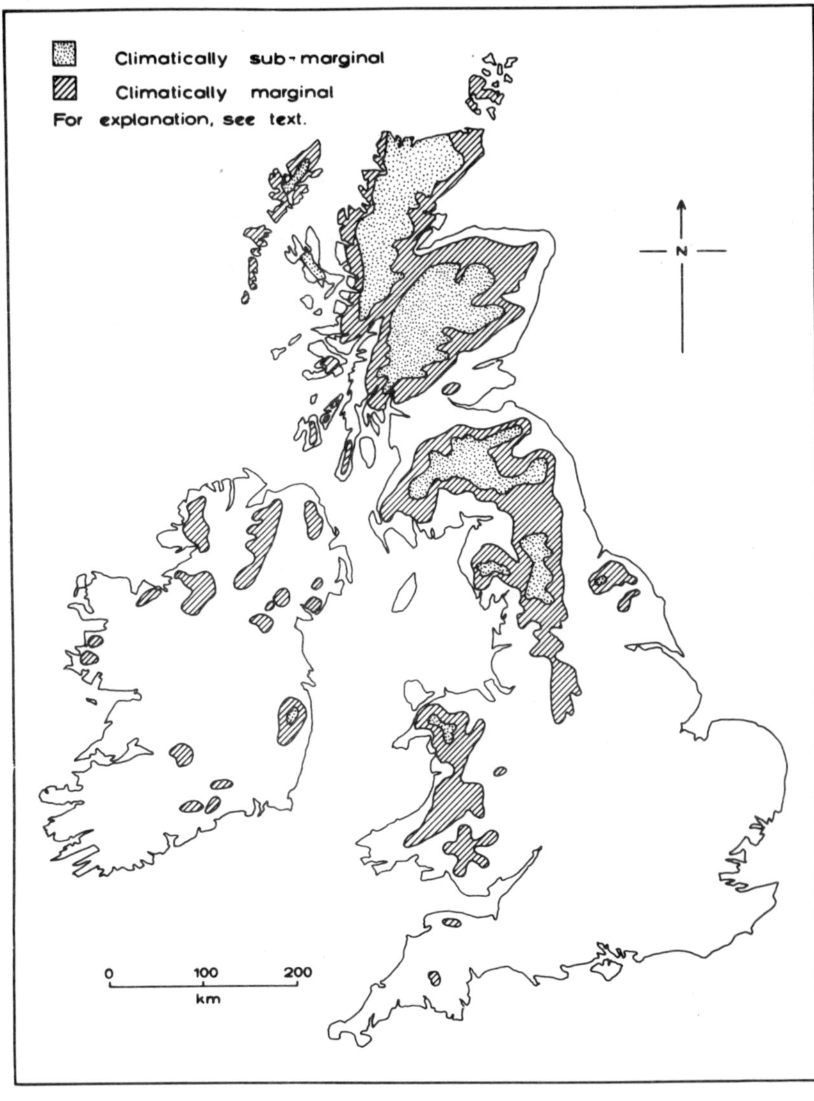

Figure 5 - Climatically marginal land in the British Isles. After Parry (1978).

windspeed, accumulated temperature and "end-of-summer" water surplus (Parry, 1975). Minimum levels of these parameters, first, for maturation of the oats crop and, secondly, for commercial cereal cultivation enable the delineation of a zone that is "climatically marginal" for oats cropping (Fig. 5). We have thus identified a "laboratory" in which the impact of climatic variability on oats farming is enhanced.

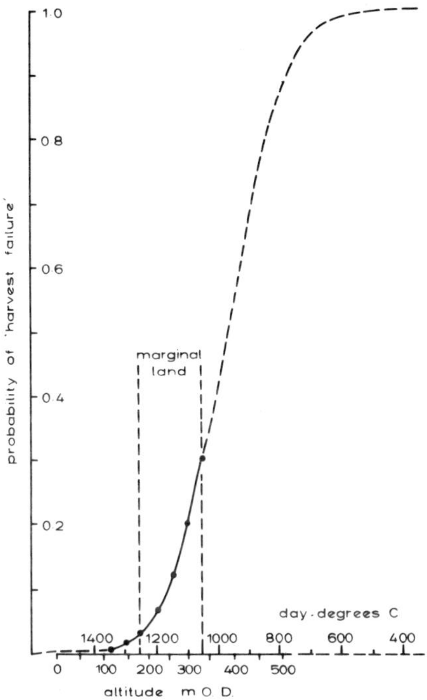

Figure 6 - Probability of 'harvest failure' in south-east Scotland. After Parry (1978).

Weather, uncertainty and farming decisions

An important link between land abandonment and climatic change is the farmer's decision. In the past, this was based on the farmer's perception of the viability of cropping at a particular location, particularly of the likelihood (i.e. the probability) of achieving a minimum yield required for survival of the farming household from year to year or of achieving an average yield that would be considered a tolerable return to effort and investment. In marginal areas of semicommercial or subsistence farming, the probability of a minimum yield (or, conversely, of failure of the harvest) was probably the more important.

This conclusion has great significance for upland maritime areas because the probability of accumulated temperatures falling below a critical level required for the ripening of oats (i.e. the probability of crop failure) increases quasi-exponentially, not linearly, with elevation (Parry, 1976). Figure 6 illustrates the curve of probability of harvest failure (i.e. the probability of occurrence of a growing season of 970 day-degrees or less above 4.4°C) for the Lammermuir Hills in south-east Scotland. At the limit of cultivation in southern Scotland (c.300m O.D.) the frequency is about 1 in 5 and doubles with a 50-metre increase in elevation. In this region the lapse rate of temperature is about 1°C per 148 metres or about 0.3°C per 50 metres. A decline in annual temperature of only 0.3°C could thus lead to a doubling in the frequency of harvest failure at 300m O.D.

Changes of agro-climate

We may conclude from the foregoing discussion that, in order to evaluate the impact of climate change on early oats cultivation in upland Britain:

1) changes in climate need to be expressed as changes in accumulated temperature, potential water surplus (P.W.S.), and exposure;
2) changes in climate need to be expressed in terms of the probability of crop failure.

When this is done the scale of impact is apparent. If, over the period A.D. 1250-1450, mean maximum temperatures fell a little less than 1°C then summer warmth at 300m O.D. in northern England was reduced by 15 per cent, the frequency of crop failure increased seven-fold (from 1 in 20 to 1 in 3), and the frequency of two consecutive crop failures increased seventy times (from 1 in 700 to 1 in 10) (Parry, 1976).

The chronology of climate change which has been constructed by H. H. Lamb for north-west Europe for A.D. 800-1900 embraces changes in 50-year averages of temperature that exceed 1°C (Lamb, 1977). These estimates contain a quite wide margin of error and most of them are not statistically significant. Doubtless they will be improved upon in time but, at the time of writing, they remain the most accurate available. From them and from our modelling of the links between weather and oats growth, we may conclude that there have been substantial changes in the viability of upland farming over the past millenium and that the climatic "limits" to cultivation have risen and fallen with the fluctuation of potential for plant growth.

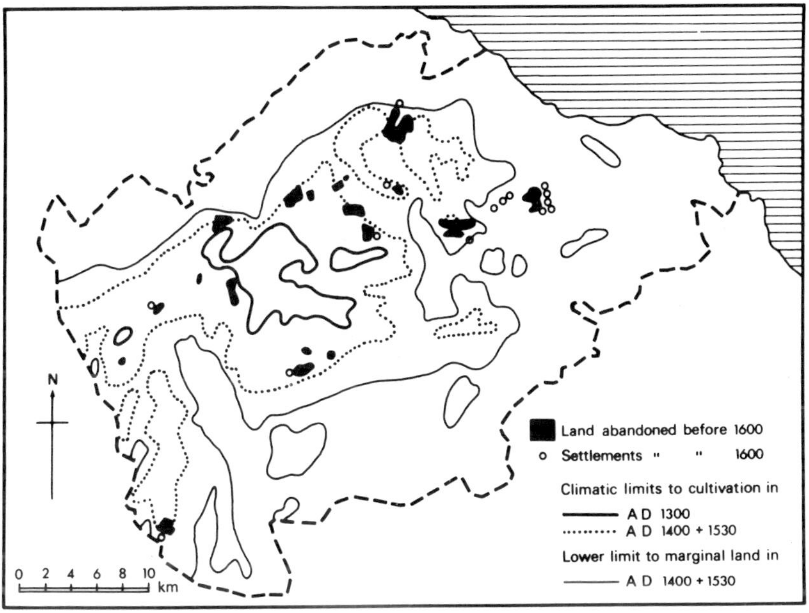

Figure 7 - Abandoned farmland and lowered climatic limits to cultivation in south-east Scotland, A.D. 1300-1600. After Parry (1975).

RETRODICTING THE CLIMATE-FARMING LINK

This analysis of the most significance sets of climate-farming processes enables a retrodiction to be made about the impact of climatic change on high-level agriculture in early modern Europe. The retrodiction is that shifts of isopleths of critical levels of the probability of crop failure, resulting from changes of climate, were correlated with areas in which most frequently occurred the permanent abandonment of farmland. This retrodiction can now be tested.

Figures 7 and 8 illustrate that there is a close temporal and spatial fit between the zone over which theoretical limits to cultivation fell over A.D. 1300-1600 and A.D. 1600-1800 and the distribution of permanently abandoned farmland. The retrodiction is thus confirmed (for further discussion of the correlation see Parry, 1978). Since the predicted space-time correlation was derived theoretically and empirically from an analysis of the processes linking climate and agriculture it is logical to assume that the observed correlation reflects some degree of causal relationship between climatic change and land abandonment.

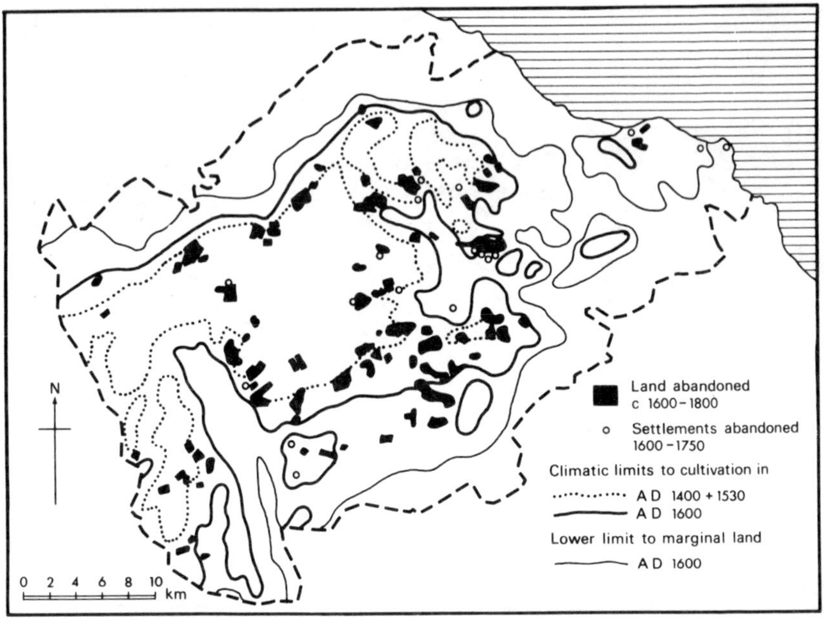

Figure 8 - Abandoned farmland and lowered climatic limits to cultivation in south-east Scotland, A.D. 1600-1750. After Parry (1975).

It is also possible to retrodict the distribution of permanent land abandonment in the same way for a similar period, A.D. 1300-1600, for the British Isles as a whole (Fig. 9). This is based solely on temperature accumulation and does not include regional and local variations in P.W.S. and exposure. Nevertheless, it points to the area where the economic historian should look for the full economic consequences of climatic cooling. This "recurrently marginal" land covers more than two million hectares, or about one-third of Britain's presently unimproved moorland; and if the pattern displayed in the Lammermuir Hills is characteristic of Britain as a whole, there may have occurred a retreat of agriculture and settlement in these areas in the late Middle Ages. This is a "retrodiction" which remains to be tested and awaits the attention of the economic historian.

CONCLUSION

The task of evaluating the impact of climate change on the path of economic history is not an easy one. It has not been made more easy by a failure of historians to grasp the complexities of climatic processes and the failure of climatologists to recognise the limits placed by

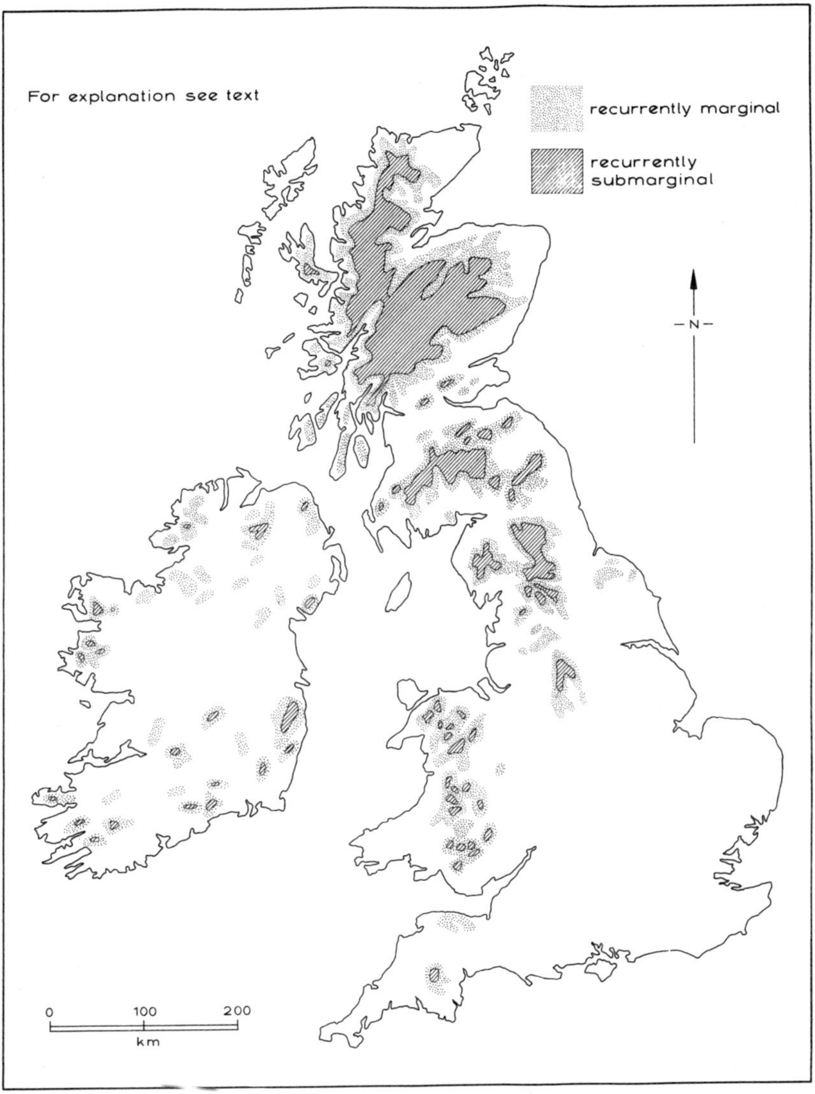

Figure 9 - 'Retrodicted' shift of cultivation limits in the British Isles, A.D. 1300-1600. The 'recurrently marginal' zone is that over which theoretical climatic limits to successful cereal cultivation have shifted during this period. After Parry (1978).

"history" on their traditional research strategies. Nevertheless, a solution to the problem may be found in a retrodictive strategy that evaluates the causality behind the coincidence of climatic and economic events in time and space by testing predicted "coincidences" derived from careful modelling of the processes linking weather and agriculture.

Finally, a note of caution is due. This discussion has focussed on only one of a number of explanations for farmland abandonment in early modern times. Other explanations suggested in Figure 1, such as errors of farming judgement, and changes in farming objectives and farming systems, remain to be tested. I hope that, by electing to draw upon the issue of climate change and land abandonment as an illustration of a promising research strategy, I have not inadvertently led the reader to assume that climate change was a "major" factor behind agricultural change. Such a misapprehension would be akin to inferring that Frazer was arguing in *The Golden Bough* for the primacy of tree-worship in the evolution of religion. Adapting Sir James Frazer's words (1922, Preface), I shall add an explicit disclaimer: that I am so far from regarding climatic change as of supreme importance to agricultural change that I consider it altogether subordinate to other factors [in particular to social and economic forces for change (Parry, 1978)] and I thus hope that I shall not be taxed with embracing a notion which I look upon not merely as false, but as preposterous and absurd.

REFERENCES

Beresford, M. W. (1971) A review of archaeological research (to 1968). *In Deserted medieval villages*, ed. M. W. Beresford and J. G. Hurst. 21-144, Lutterworth Press, London.

Brooks, C. E. P. (1949) *Climate through the ages.* 2nd ed., London.

Bryson, R. A., Lamb, H. H. and Donley, D. L. (1974) Drought and the decline of the Mycenae. *Antiquity*, 48, 46-52.

Frazer, J. G. (1922) *The Golden Bough.* Macmillan, London.

Lamb, H. H. (1977) *Climate: present, past and future*, Volume 2 *Climatic History and the Future.* Methuen, London.

Le Roy Ladurie, E. (1972) *Times of feast, times of famine.* Allen and Unwin, London.

Nagel, E. (1961) *The structure of science.* London.

Parry, M. L. (1975) Secular climatic change and marginal land. *Transactions, Institute of British Geographers*, 64, 1-13.

Parry, M. L. (1976) The significance of the variability of summer warmth in upland Britain. *Weather*, 31, 212-17.

Parry, M. L. (1978) *Climatic change, agriculture and settlement.* 144, Dawson, Folkestone.

Postan, M. M. (1975) *The medieval economy and society.* 2nd ed., Penguin, Harmondsworth.

HUMAN RESPONSE TO SHORT- AND LONG-TERM CLIMATIC FLUCTUATIONS: THE EXAMPLE OF EARLY SCOTLAND.

Ian Whyte
University of Lancaster

Previous research on the relationships between climate and human societies has concentrated on the effects of secular climatic changes on communities and not the responses of these communities to such effects. This paper analyses the responses of rural societies in early-modern Scotland to short-term weather events and argues that these hold the key to the understanding of responses to long-term climatic changes.

Previous research on the relationships between climatic change and man has been concerned with two main themes. Firstly, climatic historians have investigated historical sources for evidence of human activities and observations of natural phenomena in order to demonstrate that climatic fluctuations occurred in the past. Secondly, workers from a variety of disciplines have attempted to assess the effects of climatic episodes on human societies. Both these lines of research have often gone hand in hand using the same source material. As Parry (1975) has stressed, it is in marginal environments and their communities that the effects of secular climatic deterioration were most pronounced. In Britain such areas include clay lowlands where increases in precipitation may have made drainage and cultivation more difficult (Beresford, 1975) and coastal areas where more frequent storms may have threatened sea defences (Brandon, 1971). The uplands of northern and western Britain, however, contain the greatest area of marginal land vulnerable to detrimental changes in temperature, precipitation and other climatic parameters.

While a number of studies have been made of the effects of short-term climatic variations on past communities and some efforts have been made to assess the impact of long-term changes, there is a need for more detailed research in this field. A recent review has pointed to the general failure of such studies to develop sophisticated models of causation in investigating the impact of climate upon man in the past (de Vries, 1980, p.602). It is important at this point to distingish between the effects of short and long-term climatic episodes on human societies and the responses of societies to these effects. Behavioural geographers have studied present-day human responses to natural hazards including floods, storms and droughts in developed and underdeveloped societes (Kates, 1976). This research has been exclusively concerned with reactions to short-term weather conditions. The interpretation of human responses to long-term climatic changes has not been widely attempted. Indeed, there have been comparatively few studies of the reactions of past communities to short-term weather events in order to determine the ways in which they might have differed from those of modern societies due to contrasting social, cultural and economic backgrounds (e.g. Hoskins, 1964, 1968; Walton, 1952; Wood 1965).

There are difficulties in applying to past situations the methods which have been employed in a modern context by behavioural geographers. The time element rules out the use of psychological interview techniques which test individual responses to situations and phenomena. The use of methods such as content anaysis are only appropriate to certain categories of documentary source. Nevertheless, some attempt may be made. This paper examines some aspects of the responses to both short and long-term climatic fluctations of a particular early-modern rural society.

RESPONSES TO SHORT-TERM WEATHER CONDITIONS

In attempting to analyse the reactions of rural communities in early-modern Scotland to climatic influences it is appropriate to begin by examining responses to short-term weather fluctuations, or "adjustments" as they have been termed in studies of modern situations, as these are likely to be comparatively easy to identify. Attention will then be turned to long-term climatic changes. In both cases the nature of the farming enterprise and the character of the local and regional economy should be carefully considered when analysing the options which were available to individual farmers.

The impact of adverse weather conditions on rural communities at this time has usually been studied in terms of their effects on cereal production, which have normally been summarised in the quality of the harvest (Hoskins, 1964, 1968; Parry, 1975). This emphasis is understandable given that the diet of most European peasant societies in medieval and early-modern times was grain-based. In addition, as Parry (1975, 1976) has shown, traces of former cultivation in marginal areas can be readily identified from aerial photographs and ground survey. Nevertheless, as de Vries (1980, p.602) has suggested, in the context of post-medieval economies with well-developed trade and marketing systems and complex social structures, it is simplistic to think in terms of a straightforward model of causation in which severe weather conditions produced harvest failure which automatically led to widespread famine. A by-product of this preoccupation with harvest conditions has been the implicit tendency to consider the fortunes of agriculture in upland areas in terms of subsistence cereal production and to ignore the effects of climatic deterioration on livestock husbandry.

Livestock provided the main source of income for the bulk of upland farmers in Britain from medieval times onwards. The problems of grain supply to interior areas with pre-industrial marketing structures and transport technology frequently resulted in the addition of a subsistence arable element to predominantly pastoral enterprises but commercialization had proceeded sufficiently far to allow considerable regional specialization with interchanges via market centres located along the upland/lowland fringe. In Southern Scotland the Cistercians had developed commercial sheep rearing in the thirteenth century (Franklin, 1952). Large scale sheep farming was encroaching on royal hunting forests like Ettrick during the fifteenth century (Gilbert, 1979, 259-61). The pacification of the Borders after the Union of the Crowns in 1603 led to an expansion of sheep production in the eastern and central Southern Uplands (Register of thr Privy Council of Scotland, (R.P.C.), 1623 iv 416). Raw wool was one of Scotland's major exports in the seventeenth century while woollen cloth from the Borders also figures prominently in customs records (Smout, 1963). Increasing contact with England, particularly the growing demands of the London market, encouraged the rearing of store cattle in Galloway and parts of the Highlands in the later seventeenth century. Within an economy which was gradually becoming more commercialized upland farmers would have felt the effects of adverse short-term weather conditions more severely in terms of livestock mortality than in ruined crops as it was livestock which provided the money which paid their rents and ultimately provided them with a livelihood.

To highlight the impact of short-term climatic conditions on pastoral farmers it is useful to focus upon one severe weather episode, the crisis of the 1690s, and to analyse its effects upon a particular upland community in contrast to those experienced by lowland arable farmers. The discussion assumes that in dealing with upland settlement in early-modern Scotland we are considering farms which were leased by tenants from landed proprietors and which were integrated to varying extents into a wider regional and national economy. This is the kind of enterprise for which documentary material in the form of reports, accounts, leases and rentals is most abundant. On the other hand, where settlement margins were characterised by unauthorised squatter colonization different responses might have occurred from cultivators who were weakly integrated into the market system, more oriented towards subsistence cereal production, and possibly more vulnerable to severe weather conditions. The existence of squatter encroachments at the fringes of cultivation is sometimes suggested in the boundary perambulations of commonties or common pastures in joint ownership (Parry, 1973, p.218) and this topic requires further investigation.

The later 1690s were years of severe weather and famine throughout Europe and Lamb (1977, p.471) has viewed it as the nadir of the Little Ice Age. In Scotland the period went down in folk history as the Seven Ill Years. The work of Flinn (1976), Mitchison (1965) and Smout (1963) has suggested that population loss by mortality, emigration and a fall in the birth rate over Scotland as a whole was less drastic than tradition asserts - perhaps in the region of 15 per cent (Flinn 1976) - and that the years of high rain prices were restricted to four or five years in most areas (Smout, 1963). An early study by Walton (1952) indicated that the severity of the famine in the North East of Scotland was much greater in interior pastoral areas than in the coastal lowlands. Nevertheless, the nature of the weather conditions involved, their effects on farming enterprises, and the responses of tenant farmers and their landlords have not been studied in detail.

Good indications of the impact of the severe weather conditions of the later 1690s on pastoral farming come from the chamberlain's reports on the Buccleuch estates which covered a wide area of the Central Borders (Scottish Record Office (S.R.O.), G.D. 224 935 2,3). The reports provide sufficient detail to allow a broad chronology of weather conditions and their effects to be established. In this part of Scotland at least farmers were experiencing difficulties a decade or so before the traditional onset of hard weather in the later 1690s. The Duke of Buccleuch petitioned the Scottish Privy Council in 1675 referring to a "great and extraordinary storm" the previous year "whereby the greatest part of the....cattle belonging to their tenants were destroyed" and claiming that as a result "a very considerable part of (the) estate remains ... waste and unpossessed" (R.P.C., iv 1675 426). There are indications that the decade from 1684 to 1694 was unusually wet throughout the Borders. In 1684 and 1685 rent arrears were running at high levels in Ettrick Forest and Teviotdale with some tenants owing two entire years' rent. Some farms such as Altrieve near St. Mary's Loch were untenanted due to flood damage which in this case had destroyed five bolls sowing of arable land - about five acres.

The years 1686 and 1687 appear to have been less eventful as regards weather conditions but in 1688 the harvest was late and much of the crop was lost by shaking and rotting due to heavy rain. The winter of 1688-9 was hard with heavy losses of sheep. Whether this was due to a lack of fodder after the previous poor summer or the result of a delayed start to spring growth is not clear. The harvest of 1689 was again a bad one and conditions for cereal cultivation in the lower parts of the Border dales had become so precarious that much land in Teviotdale went out of cultivation. Heavy rain caused flood damage to hayfields and arable land in Ettrick Forest, Eskdale, Liddesdale and Teviotdale. The unfortunate tenant of Altrieve had land worth ten bolls sowing of oats - about ten acres - and two days' mowing of hay destroyed. Loss of winter fodder by the effects of rain on the hay crop may have contributed to the continuing heavy livestock mortality but in 1690, another bad year for storm and rain, for flooding and crop damage, tenants were complaining of heavy losses of sheep due to "traik". This disease may be identified as the effects of sheep maggot fly, a pest which lays eggs in the fleeces of sheep especially when they are constantly kept moist through prolonged rain and contact with wet vegetation (Robinson, 1977, p.454). The summer of 1691 brought a drought which prevented water mills from operating but the respite from wet conditions was brief. Between 1692 and 1693 several tenants petitioned for abatements of rent due to high mortality of sheep through "rot" or sheep liver fluke, a parasite which is particularly favoured by wet summer conditions (Robinson 1977, p.548).

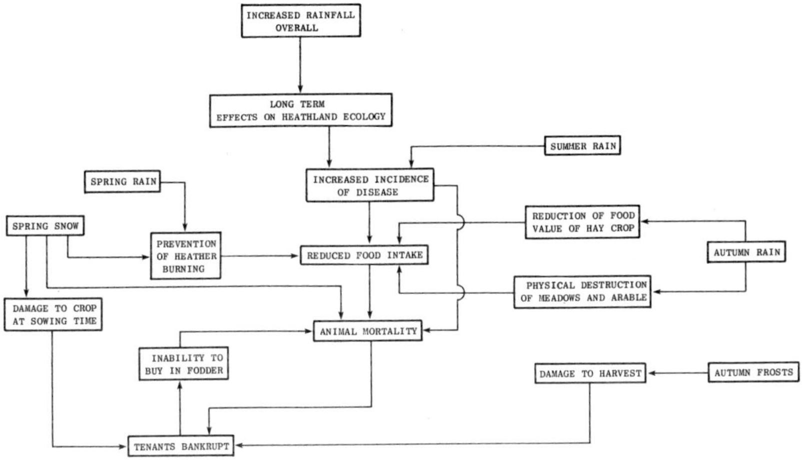

Figure 1 - Detrimental effects of a run of wet years on sheep farming in the Borders in the late seventeenth century.

In order to illustrate some of the influences which led to livestock mortality it is appropriate to summarize the known and probable detrimental effects of a run of wet years on sheep farming in this area during the late seventeenth century (Fig. 1). Livestock production and farmers' profits would have been reduced by a combination of influences. Continuous heavy rain would have reduced the time that sheep spent grazing and if this was prolonged their food intake and weight gain could have been reduced, weakening their resistance to disease and severe weather conditions (Spalding, 1965, p.203). Wet summer conditions favour many sheep diseases, especially liver fluke and sheep maggot fly but also sheep ticks and foot rot (Robinson, 1977, p.418, 641). Wet autumn conditions would have made haymaking more difficult and would have reduced the food value of the hay through loss of digestible carbohydrates due to increasing respiration, by the leaching of soluable constituents, and by moulding (Robinson, 1977, p.548). Flooding due to heavy rain physically destroyed valley-bottom hayfields. The resulting lack of winter fodder would have been especially critical if late, cold springs had delayed the onset of the new growth of grass in April or May, while reduced cereal crops would have meant that tenant farmers were forced to buy in grain for bread which they would normally have produced themselves, thus reducing their capital reserves.

Another possible effect, though one which is much harder to evaluate, was an adverse long-term change in the ecology of the upland pastures on which the sheep were dependent. In the late 1680s many tenants in Teviotdale were complaining about a lack of heather. The tenant of Dowislees stated in a petition that his ground was bare of heather which had formerly provided both food and shelter for his flocks and that as a result the carrying capacity of his farm had been reduced by a third. The onset of wetter conditions might, in some limited areas which were poorly drained, have led to the replacement of *Calluna vulgaris* by species such as *Erica tetralix*, *Eriophorum sp.* or *Molinia cerulea* which were better suited to wetter conditions but which were less nutritious. Such a problem, if it existed, would only have been of limited importance. A more serious difficulty may have been the combination of wetter conditions with over-stocking or poor grazing management which may have reduced heather cover. Alternatively, normal grazing practices may have been disrupted. Wet springs in particular may have made the burning of heather more difficult and less effective and this may have reduced the amount of nutritious young heather which was available. These suggestions are, however, extremely tentative and they illustrate the difficulty of trying to identify the effects of influences such as systems of grazing management which could, on one hand, have operated independently of climatic factors to modulate or increase their effects and which, on the other, may have been altered as a direct response to deteriorating conditions.

The picture then, is of a run up of at least a decade of difficulty to the grim years of the later 1690s. The winter of 1694-5 was severe, as was the winter of 1695-6. A combination of shortage of fodder and heavy snow in the spring of 1696 led to further heavy animal mortality and the annual meeting of tenants at Hawick in April to renew their leases was abandoned as many of their horses were too weak to undertake the journey. Early frosts in September 1696 resulted in the almost total destruction of the grain crop while late frosts in the spring of 1697 delayed sowing dangerously late. Summer floods in 1697 washed away crops on many farms or buried them under sediments while early frosts blackened any crops which had survived. The continuation of cold, wet conditions in 1698 and 1699 again resulted in the abandonment of cultivation on many farms.

There was undoubtedly great hardship among the tenantry although the level of mortality is not known. Particularly after the crop failures of 1696 and 1697 tenants in the lower parts of Teviotdale who were more dependent upon arable farming and who paid their rents substantially in grain were unable to pay anything and were selling their effects to buy seed corn and to keep their families from starving. Where crops could be harvested at all four, five, or six bolls of oats only produced one boll of meal which was of such poor quality that "none but in gret necessity can make use of". Where the chamberlains were able to gather in grain rents at all they found their quality so poor that the grain could not be marketed for either bread or seed. Likewise tenants on pastoral farms had no money with which to pay their cash rents because of the severe reduction of their flocks. The chamberlains petitioned the estate commissioners in Edinburgh that despite the rapidly mounting arrears tenants should not be evicted, particularly those who had paid their rents regularly in previous years.

Recovery from such a run of harsh seasons would have taken longer for farmers in upland pastoral areas than for the tenants of lowland arable holdings. The latter could have obtained seed corn with which to recommence cultivation comparatively easily as long as their landlords were prepared to grant them credit as international trade was geared to the transport of grain. It would have taken far longer to build up flocks and herds to their old levels and pastoral estates must have experienced a shortfall of rents for several years after the disaster. The picture for the Scottish Borders then is not the one of a short run of exceptionally severe weather conditions between 1695 and 1699 as has been suggested for Scotland as a whole, but one of long-term difficulty extending over at least 15 years during which conditions were consistently unfavourable to sheeep and cattle rearing.

By contrast the lowland arable areas of eastern Scotland do not seem to have suffered such severe dislocation for as long a period. Such areas were less vulnerable to the effects of orographic precipitation, flooding, lower temperatures and reduced sunshine. At Dirleton in East Lothian, while the quantity of grain exported from the estate fell by a third between the harvests of 1697 and 1699, the sale of grain to Edinburgh merchants was able to continue (S.R.O., G.D. 6 1542, 1544). The lists of arrears show that some tenants were unable to pay their grain in full but at their peak the arrears amounted to only a fifth of the annual rental of the estate with only half the tenants being in debt for substantial amounts. Similar patterns can be discerned on other estates around the Firth of Forth (Whyte, 1979, p.248). Records of parish poor law administration indicate that within the region there were marked differences in mortality between parishes which were relatively generous in the disbursement of funds to the poor and those which were more tight-fisted, (Mitchison, 1974) but despite high grain prices the indications are that the upper levels of peasant society, the tenant farmers, suffered comparatively little.

Arable areas north of the Tay appear to have been affected more strongly but again not on the same scale as in the Borders. On the Panmure estates in Angus the sale of grain to merchants in Dundee, Edinburgh and Glasgow continued at a reduced rate. Nevertheless, rent arrears among tenants in the barony of Panmure, which had remained at under £2,500 for most of the previous three decades, suddenly rocketed after the harvest 1696 and reached over £20,000 Scots after the crop of 1699 (S.R.O., G.D. 45 18). The condition of the tenantry may not have been as grim as these figures might superficially suggest. The arrears are inflated by the system of accounting in which tenants who could not pay their grain rents in a year of shortage were debited at the high market price for that year. Thus, after a poor harvest a modest shortfall of grain could have produced large money arrears which would have taken several season at low prices to write off.

ADJUSTMENT AND ADAPTATION:

Coping strategies in the short and long-term

Having considered the effects of a period of severe weather conditions on farmers in arable and pastoral areas it is possible to move towards a more general model of human responses to such conditions. Figure 2 shows various consequences and the strategies which could be adopted to cope with them as well as the eventual outcome of the actions taken. The outcome for any holding or farm can be summarized in the short term as whether or not its tenants continued farming and in the long term as whether or not the farm itself remained viable. Short-term responses will be examined first as these hold the key to the understanding of long-term changes.

In the context of seventeenth-century Scotland it is inappropriate to apply the sophisticated models of decision theory which have been used to explain strategies adopted by modern farmers. The options of a Scottish tenant farmer at this time in terms of modifying his enterprise in both the short and the long term were limited by several constraints. The weight of custom and tradition, restrictions imposed by systems of co-tenancy, lack of knowledge of alternative systems of husbandry, under-capitalization, and limitations imposed by rent and marketing structures all reduced the scope for adjustment or adaptation. It may have been possible to vary the balance of an enterprise slightly between arable and pastoral farming or between sheep and cattle production within the livestock sector but the scope for manoeuvre was restricted. Essentially the outcome depended upon the interaction of the responses of two people; that of the tenant based on his farming skills and capital reserves, and that of the proprietor influenced by his appreciation both of the wider economic situation and his understanding of conditions on his own estate.

The question of whether Scottish tenant farmers at this period had sufficient capital reserves to tide them over one or two bad seasons without requiring aid from their landlords has yet to be answered. A study of the testaments and inventories of people who can be identified from estate records as tenant farmers would shed some light on this problem. For the moment it will be assumed that their reserves were generally limited. In a crisis brought about by severe weather conditions their problems would have revolved around balancing the conflicting demands of paying their rents, maintaining their families and allocating sufficient resources to replace losses and continue farming. Inevitably the logical choice was to leave rents unpaid. A proprietor's reaction to a situation in which most, if not all, of his tenants were in arrears would have differed from his attitude towards individuals who could not pay due to bad management under reasonable conditions. Wholesale eviction would have achieved nothing. It was in a landlord's interest to keep his tenants on the land. On the other hand evidence from the Buccleuch estates shows that the consideration which tenants received in a crisis could depend upon their past records of payment or non-payment. Where information is available on the performance of tenants in the years preceding a crisis it should be possible to predict those who were more likely to be at risk of eviction as a result of bad management and/or bad relations with their landlord, and to compare the predictions with reality in the light of the policies adopted by the estate.

More research is needed to determine the stability or rate of turnover of tenants after short-term crises but if we move to a study of reactions to long-term climatic changes we are on even less sure ground due to the difficulty of separating responses which were related to climate from the multiplicity of other shifting parameters. Changes in weather patterns over periods spanning several decades might have left an imprint on the folk memory of farming communities but this is rarely likely to be recorded. The decision-making processes of both tenant and proprietors with regard to the viability of marginal farms which were at risk during a period of long-term climatic deterioration are hard to detect. Indeed, it is unlikely that there was a real long-term perspective on either part. A tenant's experience of success or failure on any farm is likely to have extended back no further than two generations at best - his own and his father's. While a proprietor or his factor might have had past rentals and leases to guide them it is unlikely that decisions relating to rent and tenancy conditions were made using long time series of data. In practice the reaction of proprietor and tenant to long-term climatic deterioration was probably the net result of a series of decisions taken in response to individual bad years so that the model proposed in Figure 2 can be extended from adjustments to short-term conditions to adaptations to long-term ones.

Climatic deterioration in marginal areas is more likely to have manifested itself to farmers and their landlords in terms of increasing frequencies of short-term difficulties rather than in declining average productivity over extended periods. Thus, the higher incidence of harvest failure as discussed by Parry (1975) was probably of greater significance than a gradual drop in mean yields. After each bad season and resulting short-term crisis the responses of tenant and proprietor would have been as shown in Figure 2. In the longer term a tenant might have tried to modify his enterprise. This would normally have required the landlord's approval and where documentation is adequate changes in the balance of cropping to livestock rearing should be detectable from rentals and leases as in Lowland Scotland at this time rents for arable land were normally paid in grain and rents for pasture in money. Changes in farming patterns within the pastoral sector would be harder to detect but for the earlier seventeenth century teind (tithe) returns may provide data. A proprietor's reaction to a situation where the tenant or tenants of a particular farm were frequently in arrears could have been either negative or positive. A negative response would have been to blame poor husbandry and to hold rents steady. The result of this might have been a rapid turnover of tenants who were unable to make the farm pay. Potentially this might have led to the eventual

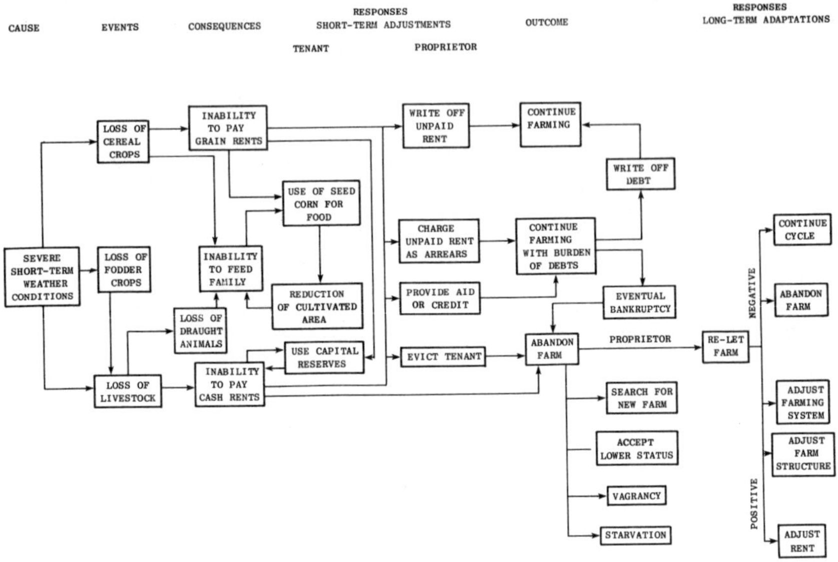

Figure 2 - Adjustments and adaptations of tenant farmers and proprietors to short- and long-term climatic changes.

abandonment of the farm. This would have been an extreme solution which may not have been widespread even at the margins of cultivation and it is worth looking at this concept more closely. The abandonment of a farmstead is not necessarily the same thing as the abandonment of a farm. In practice it is unlikely that farms were ever totally abandoned in the sense that the settlement, its cultivated lands and its grazings were no longer used at all. In most cases amalgamation with another farm is more probable, possibly combined with a change to a less intensive pattern of land use. Abandonment is thus a relative rather than an absolute concept. Landscape evidence of deserted settlement sites and cultivation ridges, while dramatic and evocative, does not allow us to identify which of a range of organizational options and decision making processes were adopted to create them.

A more positive response by a proprietor would have been to adjust the rent of farms which were in difficulty, to make allowances for declining productivity or changes in the nature of the enterprise. In either case the adjustment, in the face of climatic deterioration, would be a downward one either absolutely or relatively.

There is plenty of evidence that while rack renting occurred in seventeenth-century Scotland with holdings sometimes being leased to the highest bidder the rents of farms were often subject to prior negotiation before entry between a prospective tenant and a landlord

(Whyte, 1976, p.8). Reductions of rent occurred widely on the Gordon estates in North East Scotland in the early eighteenth century when farms which had been abandoned during the famine years were released often at half their previous rents (S.R.O., G.D. 44 51 77, 167). On the Buccleuch estates too, holdings which had been subject to flood damage during the 1690s had their rents reduced.

CONCLUSION

In attempting to probe problems of human responses to climatic conditions within a particular rural society and period this study has posed more questions than it has answered. The model which has been proposed is specific to a particular context. While further research might allow it to be corrected and refined within this context it would require modification in order to be applicable to other rural societies at different periods. Nevertheless, it has demonstrated the complex nature of the reactions involved depending as they did upon the interplay of social and economic as well as environmental factors all of which would have varied not only temporally but also spatially within any period.

Landscape evidence is invaluable in demonstrating that fluctuations of the margins of settlement occurred in the past. By establishing a chronology of such fluctuations it is possible to infer something of the processes which produced them. As this study has suggested, however, even the concept of abandonment is not a straightforward one. More information is needed about the structure and operation of rural communities to gain insights into the processes which created the features which can be identified in the present landscape. There is plenty of scope for further research on the behavioural responses of individuals and communities in the past in the face of natural conditions. Three lines of enquiry would be worthy of investigation. Firstly, direct indications of the decision-making processes which operated can sometimes be gained though this is more likely to be forthcoming for proprietors than for the farmers themselves. Secondly, the results of the decisions which were taken can be used as a means of working back towards the decisions themselves. For example, changing spatial patterns of tenancy, farm structures and levels of arrears, especially between lowland core areas and the upland fringes of large well-documented estates, could shed light on the responses of both landlords and tenants though it must be admitted that it would be difficult to separate the effects of climatic changes from those of other variables. Thirdly, information could be sought on conditions which influenced decision-making such as the extent of farmers' capital reserves and the degree of freedom available to them for modifying their farming systems. Documentary sources capable of providing material of this kind are not likely to exist for England much before the sixteenth century and certainly not for Scotland before the later seventeenth century. This time limit nevertheless allows some notable short-term crises such as that of 1623 in Northern England or the later 1690s in Scotland to be considered. Ideally, in the case of well-documented examples, it should be possible to link landscape features with cartographic and other evidence to provide a chronology of change and then to use documentary material to examine the processes which created the observed patterns.

REFERENCES

Beresford, G. (1975) *The medieval clay-land village. Excavations at Goltho and Barton Blount*. London.

Brandon, P. F. (1971) Late medieval weather in Sussex and its agricultural significance. *Transactions, Institute of British Geographers*, 54, 1-18.

Flinn, M. (1976) *Scottish population history*. Cambridge.

Franklin, T. B. (1952) *A history of Scottish farming*. London.

Gilbert, J. M. (1979) *Hunting and hunting reserves in medieval Scotland*. Edinburgh.

Hoskins, W. G. (1964) Harvest fluctuations and English economic history 1480-1619. *Agricultural History Review*, 12, 28-47.

Hoskins, W. G. (1968) Harvest fluctuations and English economic history 1620-1759. *Agricultural History Review*, 16, 15-31.

Kates, R. W. (1976) Experiencing the environment as hazard. In S. Warner, S. Cohen, and B. Eaplan, (eds.) *Experiencing the environment*. New York.

Lamb, H. H. (1977) *Climate present, past and future*, Volume 2: *Climatic history and the future*. London.

Mitchison, R. (1965) The movements of Scottish grain prices in the seventeenth and eighteenth centuries. *Economic History Review*, 2nd ser., 18, 278-91.

Mitchison, R. (1974) The making of the old Scottish poor law. *Past and Present*, 63, 58-93.

Parry, M. L. (1973) *Changes in the upper limit of cultivation in South East Scotland 1600-1900*. Unpublished Ph.D. thesis, Univ. of Edinburgh.

Parry, M. L. (1975) Secular climatic change and marginal agriculture. *Transactions, Institute of British Geographers*, 64, 1-14.

Parry, M. L. (1976) The abandonment of upland settlement in Southern Scotland. *Scottish Geographical Magazine*, 92, 50-60.

R.P.C. *Register of the Privy Council of Scotland*. 2nd ser., 1625-60. Edinburgh 1899-1908; 3rd ser., 1661, Edinburgh, 1907-70.

Robinson, D. H. (1977) *Elements of agriculture*. London.

Smout, T. C. (1963) *Scottish trade on the eve of the Union 1660-1707*. Edinburgh.

Spalding, C. R. W. (1965) *Sheep production and grazing management*. London.

S.R.O. *Scottish Record Office*. H.M. Register House, Edinburgh.

Symon, J. A. (1959) *Scottish farming, past and present*. Edinburgh.

de Vries, J. (1980) Measuring the impact of climate on history" the search for appropriate methodologies. *Journal of Interdisciplinary History*, 10, 599-630.

Walton, K. (1952) Climate and famines in North East Scotland. *Scottish Geographical Magazine*, 68, 13-21.

Whyte, I. D. (1979a) *Agriculture and rural society in seventeenth-century Scotland*. Edinburgh.

Whyte, I. D. (1979b) Written leases and their impact on Scottish agriculture in the seventeenth century. *Agricultural History Review*, 27, 1-9.

Wood, D. J. (1965) The complicity of climate in the 1816 depression in Dumfriess-shire. *Scottish Geographical Magazine*, 81, 5-17.

CLIMATIC CHANGE AND ITS EFFECT UPON THE SETTLEMENT AND DESERTION OF MEDIEVAL VILLAGES IN BRITAIN

Guy Beresford
North Lincolnshire Archaeological Unit

In the British Isles the effect of climatic change would be felt mainly in areas where the land could be concidered marginal owing to its altitude, exposure to strong winds or to the nature of the soil. In extreme marginal places the smallest change could have a dramatic effect upon the ecology of its community.

During the last fifty years climatologists have been piecing together the history of the climate since the end of the last Ice Age about 10,000 years ago. Research over the past twenty has greatly extended our knowledge of the nature and chronology of the climatic change in N.W. Europe (Lamb, 1972-7). The research indicates that a post glacial climatic optimum extended from about 4,000 B.C. to about 2,000 B.C. when world temperatures averaged possibly some 2-3°C higher than they are today. The decline from the optimum was at first gradual then abrupt circa 500 B.C. There then followed a cool rainy period which eventually gave way to the second climatic optimum between A.D. 400 and 1200 with peak years probably occurring between 800 and 1000 when world temperatures probably averaged some 1-2°C higher than at present. A decline set in again and the period between 1200 and 1400 contained years of remarkable climatic instability in Western Europe with great floods and droughts, notably mild and notably severe winters. The effect of this variation of the climate would be felt mainly in areas where the land could be considered marginal owing to its altitude, exposure to strong winds or to the nature of the soil. In extreme marginal places the smallest change in the climate could have a dramatic effect upon the ecology of its community. Although the effects of climatic variation could be harsh in such sites the effect upon the long-term national economy would have been slight owing to the small areas involved. Recent field-work and archaeological excavation on both highland and lowland sites, where the land can be said to have been marginal before the advent of modern agrarian technology, has demonstrated that the variation of the climate had a far greater effect upon their ecology than some historians and others unfamiliar with agriculture are, at present, prepared to accept.

DARTMOOR

The long term climatic variation (Beresford, 1979) is clearly reflected in the corn growing uplands of Dartmoor, Devon, where the upper limit of land capable of bearing crops is dependent upon the length of the growing season - that is the number of growing months with a mean temperature of over 10°C. Dartmoor is the largest of the great granite bosses which form the high moorland in South-West England and covers an area of some 775 square kilometres. The upland area is everywhere over 213m above sea level and rises to over 620m on Highwillais in the north. Much of the open moorland is relatively level and could be cultivated in favourable climatic conditions although the spread of granite boulders could make cultivation in places difficult. The average rainfall today varies from some 1140-1650mm in the east to 1650mm in the west (ibid p.102).

During the years of the post glacial period of climatic optimum, circa 4000-2000 B.C., Dartmoor was exploited by Mesolithic man, (Radford, 1952; Greig and Rankine, 1956) but there is no archaeological evidence to suggest that the moor was widely settled before the second millenium B.C. (Denford, 1975). Prolific remains of Bronze Age occupation is attested by numerous hut circles, barrows, lynchets and enclosures showing that the moor was widely occupied and cultivated up to 455m. The decline from the climatic optimum, at first gradual and then abrupt about 500 B.C., led to the complete desertion of the moor lying above 300m in favour of valley settlement (Staines, 1979). Although many sites have been excavated on Dartmoor, there is still no evidence to suggest that the land above 300m was occupied during the cool rainy period during the middle and late Iron Age or in the Roman and early post Roman periods. Nevertheless, it may be assumed that this higher land continued to be exploited for its summer grazing by those settled in the valley bottoms. The results of recent extensive excavations at Houndtor and Hutholes by the late Mrs. Minter show that settlers were beginning to return to higher parts of the moor in the years A.D. 800-1000 a period which was (as far as we know) characterised by more frequently warm and dry summers than the preceding centuries. During this period many of the abandoned prehistoric settlements were reoccupied and cultivated up to the altitude of 395m that is, some 60m lower than the upper limit of the Bronze Age, when temperatures were possibly slightly warmer (Lamb, 1977). The sites of many of the deserted prehistoric settlements with their well watered grass and heather moors would certainly have attracted settlers in the favourable climatic conditions in the early medieval period. The early fields, already partly enclosed and cleared of boulders, must have proved an additional attraction. The excavations of 12th and 13th century settlements on the moor (Beresford, 1979, 144-5) indicate that they were prosperous communities and that the standard of living of the peasantry was similar to that of those living in lowland villages. The farmsteads about 300m were gradually abandoned during the late 13th and 14th centuries which, Lamb suggests (1977), was characterised by more frequently cool and moist summers. One hundred and ten deserted medieval settlements have been recorded on Dartmoor, the majority of which appear to be homesteads with one, two or three buildings, but hamlets with up to thirteen buildings have been found at Blackaton, Widecombe-in-the-Moor. Of these 27 lie between 210m and 275m, 22 between 275m and 305m and 61 over 305m (Linehan, 1966, Table 11). Similar instances of abandonment and reoccupation of farmsteads have been recorded elsewhere; a farm in central Norway at the upper limit of agriculture has been abandoned three times since it was first established around A.D. 400, each time during a period of deteriorating climate and left for one or more centuries before being reoccupied when the climate improved (Lamb, 1978, 13).

THE DESERTED MEDIEVAL VILLAGE OF HOUNDTOR

The deserted medieval village of Houndtor (Beresford, 1979, 102-4) lies midway between Hound Tor and Greator Rocks (SX 745789) in the southwestern corner of the parish of Manaton 350m above sea level, protected from the prevailing wind by Houndtor Down, which rises some 45m above the settlement. The fields associated with the village lay between 335m and 395m. The surrounding area had been cultivated in the Bronze Age, the early exploitation of the site being attested by the hut-circles, pounds and the lynchets upon which the village was superimposed. The excavation of the medieval settlement by the late Mrs. Minter revealed a long sequence of occupation probably dating from about 700-800 to about 1350. Most of the stone buildings of the 13th and 14th centuries - comprising four long-houses, four smaller houses, one shed and three corn-drying barns - were superimposed upon a long sequence of turf-walled houses, indicating that, apart from some minor change in house alignment, there had been little change in the layout of the village from the date of its foundation to that of its desertion in the first half of the 14th century.

The earliest of the medieval buildings at the village site were the three sunken-floored huts, each just over 3.5m long and 2m wide (ibid. 110-124). The opposing doorways and the centrally placed hearth in the best preserved hut indicate that they were used for domestic occupation and not as subsidiary buildings associated with the turf-walled houses. The absence of severe burning on any of their hearths shows that their occupation was of short duration. Since these remains are so similar to those of the sheilings in Scotland (Curwen, 1938) and to the hafodau in Wales (Beresford and Hurst, 1971, 254) it is probable that these huts were built to be occupied by herdsmen who grazed their stock on the open moorland pastures during the summer months before the establishment of permanent settlement. Similar examples of the establishment of farmsteads and small hamlets founded on the sites of summer pastures are to be found in Wales (ibid).

The period of transhumance at Houndtor was short and permanent settlement was soon established. The excavation revealed that the houses built before the middle of the 13th century were of turf. Although the remains were sparse and few complete plans survived, sufficient remained of the buildings to provide much information about their construction and development. They ranged in size from the smallest, 7.6m long and 3m wide, to those which were over 10.6m long and 4.3m wide. The walls were built of turf-grass and earth and were approximately 1.2m thick. Their inner faces were wattled, similar to those of the house seen by James Boswell and Dr. Johnson in the Hebrides in the 18th century (1807, 4th ed., 125; see also Johnson 1825, p.76). The considerable size of these buildings shows that they were built for occupation throughout the year. Although these buildings were substantial there is no evidence to suggest that they were long-houses, buildings in which man and his animals lived at opposing ends under a common roof and sharing a doorway in the lateral wall.

The houses of the second structural period were substantial stone built long-houses (Beresford, 1979, 224-240), built to last for a long period of time. The use of stone for the construction of houses cannot be associated with the deterioration of the climate since recent excavations in Britain (Beresford and Hurst, 1971, 93-5) and on the Continent (Grimm, 1939) have demonstrated that in areas where stone was readily available, stone houses began to replace those of timber or other insubstantial material as early as the late 12th century.

Houses in which man and his farm animals lived under a common roof have a long tradition on the Continent dating back to the Bronze Age and many examples have been excavated in Norway, Denmark and Sweden showing that it was the characteristic house plan in Scandinavia during the Iron Age and Viking Period (Roussell, 1934). Although long-houses occur in the British Isles at Jarlshof in the Shetlands (circa 850) (Hamilton, 1956) and there is an isolated example from Mawgan Porth, Cornwall (dated circa 950) (Bruce-Milford, 1956) there is little evidence to suggest that there was widespread distribution of this type of building until the late 13th or early 14th century. The apparent dearth of long-houses of the Saxon and Norman periods might be attributable to the more temperate climate of Britain, where small numbers of cattle could always be out-wintered in the fields especially in mild winters. However, the number of cattle held by the peasantry and the increasingly adverse winter weather of the late 13th and early 14th centuries seem to have led to the widespread construction of long-houses in pastoral areas in Britain in order to protect the meadows from the treading of the animals in winter. The same period saw the construction of cattle yards in the arable areas of eastern Britain (Beresford, 1975).

AGRICULTURE ON HOUNDTOR IN THE MIDDLE AGES

The substantial bower in the lower ends of the long-houses and the well preserved remains of the open fields indicate that the community was dependent upon mixed farming for their livelihood (Beresford, 1979, 42-3). The fertility of the land was probably maintained by regular manuring of the infield and by the exploitation of the outfield for as long as it could produce crops after resting and burning. In order to have the maximum amount of land available for outfield cultivation, and to give the land the longest possible rest between croppings, every arable acre would have been utilised.

Recent research by Parry (1978) has demonstrated how long term climatic change may have made an impact on the corn growing upland districts of Britain, where the upper limit of land capable of producing crops depends on the length of the growing season. He has also demonstrated how the theoretical limit to cultivation can be seen to have risen and fallen over the past millenium according to the long term variation of the climate. For example the average levels of warmth and wetness at 320m in the Lammermuirs in Scotland in the 13th century were prevailing at elevations of only 195m in the mid-19th century. On Dartmoor, in the south of England, the limit to cultivation probably fell from 425m to 305m (Beresford, 1979, 143-5). Estimated trends show that when the summer warmth is about 1300 day-degrees C, the frequency of crop failure is probably one year in sixty; at 1150 day-degrees C, the chances of failure would be one in eight; and should summer warmth fall to 950 day-degrees C, the chances of harvest failure could be as much as one year in two.

There are many conditions directly or indirectly attributable to climatic deterioration (ibid, 142-5) which may have led to total or partial crop failure during this period, a few of which may be briefly mentioned here. The cooler damper weather would certainly lead to greater acidity of the soils on Dartmoor; cold wet springs would have led to late sowing and reduced yields and would also have provided conditions which leave cereals, particularly oats, susceptible to attacks by leather-jackets and this, itself, is often accentuated by birds uprooting the plants whilst seeking the larvae; damp summers provide the right conditions for the spread of mildew; cold autumns lead to the immaturity of crops at harvest time; strong winds and rain lead to the lodging of the cereals. The poor harvests would give poor return on the seed planted and lead, in extreme circumstances, to the use of the following year's seed for animal feed and domestic consumption.

Although there are few diseases in animals which may be directly attributed to the weather, the climatic deterioration of this period with its associated wet summers and warm winters provided conditions to assist the spread of anthrax, foot-and-mouth disease, liver fluke and pleuro-pneumonia, all probably grouped under the medieval heading of murrain, and other diseases as foot-rot, laminitis and lung-worm. Poor quality hay leads to the emaciation of stock. Shortage of winter feed leads to the unduly early turning out of the cattle in the spring which in some instances can lead to a shortage of grass and a poor crop of hay for the following winter. Poor crops of hay may also lead to the autumn slaughter of breeding stock. The cattle plague of 1319-21 and the sheep murrain of 1315-17 combined with the loss of crops created a serious agrarian crisis in many areas and if these conditions pertained at Houndtor they would have caused particular hardship owing to the altitude and exposed position of the settlement. Houndtor was not an isolated example on the moors. Extensive excavations at Hutholes, lying some 3.75 miles to the south-west, have revealed a similar sequence of occupation. That site lies just above the 1000ft contour at the point where the modern cultivated fields give way to the open moorland.

The first phase of the cooling in the 13th century was quite probably gradual and its effect may have caused little inconvenience to those farming on Dartmoor. Although there were disastrous harvests in 1315, 1316, 1339 and 1349 there were certainly years of plenty, but as average temperatures dropped, so the tendency of harvest failure on the higher parts of Dartmoor increased. The construction of corn-driers at both Houndtor, Houndtor Farmstead and at Hutholes during this period cast a little light on the difficulties encountered on the moors as the weather deteriorated. The conversion of abandoned houses for the use of barns and corn-driers at Houndtor and Hutholes shows that the desertions were gradual. As in modern agriculture, there are always some who are less well adapted to face serious difficulties following a disastrous harvest and who leave while others manage to soldier on until the next crisis or until the end of their lives. So one by one the homesteads were deserted and by the middle of the 14th century the hardships were such that the settlements above 300m were once again abandoned.

THE CLAY-LANDS OF THE EAST MIDLANDS

The present archaeological evidence suggests that the early Germanic settlers in the 5th, 6th and 7th centuries A.D. neglected the clay-lands in favour of the more easily worked lighter soils of the chalk, gravels and sands. Most of these light soils could have produced reasonable crops under the wetter conditions of the period by constantly bringing into cultivation fresh land after long periods of rest and by the use of all the available manure, but over-cultivation of the very sandy soils would very soon break down the soil structure rendering the land little better than a desert. Such conditions probably led to the desertion of the Breckland of N.W. Suffolk in the 7th century (Clarke, 1937, 90, 93 ff.). The increase of population and the failure of some of the lighter soils to stand up to continuous cultivation encouraged the exploitation of the clay-lands in the peak years of the second climatic optimum during the 9th and 10th centuries (Beresford 1976).

Much of the clay-lands is rich and, when properly managed, is capable of producing very good yields (Beresford, 1975, 50-2). Clay soils require seasonal drying and shrinkage to recondition their structures. This rarely occurs under fallow and is only occasionally found under cereal crops. The three and two field system, with frequent fallows and with little or no drainage would have created severe soil problems. Progressive compaction would tend to break down the open texture of the soil to form impervious pans thereby causing crop failure. Such deterioration of the soil could occur even in a constant climate and would be difficult to rectify without a long period under grass. Today the pans may be broken up by deep ploughing or by subsoiling.

Many of the clay-land soils retain to some extent the qualities of pure clay. They must not be worked in wet conditions since they become puddled and impervious to water, and they must not be overstocked in winter months when under grass, since treading pokes the surface of the pasture and excludes the passage of air and moisture from the ground, leading to the eventual deterioration of the foliage. Successful arable farming on the clay-lands depends upon reasonable weather. So long as the land can be ploughed in the autumn, when the ground is relatively dry, the winter frosts will break down the furrows and a fine tilth can easily be obtained in the following spring. If, however, the land is very wet in the autumn, and the fields lie unploughed until the spring, the furrows will so bake in wind and sun that a fine tilth will be impossible to obtain. Even with modern machinery, hundreds of acres were left uncultivated on the Lincolnshire clay-lands in the wet spring of 1969 for just these reasons. One may therefore envisage the difficulties which would have been encountered by a medieval peasant who had only relatively flimsy harrows and land with little or no drainage.

It has been argued that the desertion of certain medieval villages situated upon the clay-lands cannot be attributed to the rains of the 14th century because deserted and non-deserted villages are to be found side by side (Beresford and Hurst, 1977, 21). However, it is not so much the quantity of rain which affects the ground but the ability of the fallen rain to drain away without leading to defloculation of the clay, to spring baking of the furrows or to late sowing. Soil on the clay-lands, as elsewhere varies from field to field, farm to farm and parish to parish. In some instances, therefore, it is this soil variation, rather than the selectivity of rain-drops, which can account for the proximity of deserted and non-deserted villages.

GOLTHO

Recent excavations at the deserted medieval village at Goltho (Beresford, 1975, 100-2) have revealed much information on the ecology of a clay-land settlement in the east Midlands. The site lies 21m above sea level just over 14km to the east of Lincoln. The soils are mostly of the Ragdale series and are subjected to severe waterlogging and compaction and are difficult to manage even with modern machinery in adverse climatic conditions. In such areas it takes only a little extra rain combined with a few extra days when there is little or no evaporation to make the ground extremely difficult if not impossible to work. Similar conditions are to be found at most of the sites of the twenty six medieval settlements which lie within a ten kilometre radius of Goltho.

Goltho was first occupied from circa A.D. 50-250 but owing to the limited area chosen for excavation, it was not possible to determine whether there was just an isolated farmstead or whether the huts excavated were part of a larger settlement (Beresford, 1976). The excavation of the Romano-British levels revealed a sequence of three circular huts. The spread of domestic and structural debris indicate that the huts were replaced by a complex building of Roman design, the site of which lay outside the area of excavation. The middle of the 9th century saw the construction of the fortified manorial enclosure, the hall of which was similar in size to that at the royal palace at Cheddar. Throughout the Middle Ages the manor continued to be a capital messuage of considerable importance. The homesteads of the peasantry continued to increase in number and by the time of the commencement of the desertion in the middle of the 14th century there were about thirty-seven of them.

The excavation of the village site suggested that the economy of the community was dependent upon arable farming. There was no evidence that cattle were brought into the homestead enclosures until the second half of the 14th century. This was a very late development of the village, introduced as it was gradually being deserted. The cattle were housed in open yards locally known as 'crew-yards'. Excavation and field-walking after bulldozing and ploughing showed that there were three or four such yards at Goltho. These homesteads, similar to those at Barton Blount, Derbyshire, were the last to be deserted.

CONCLUSION

Medieval society was complex. There are, therefore, many circumstances which could have affected the ecology of a settlement and considerable care must always be exercised when attributing changes to climatic variation. If one settlement were to be examined in isolation no satisfactory solution could be found, but examination and comparison of evidence from both highland and lowland settlements shows how the long term variation of the climate affected the ecology and pattern of settlement on sites which could be considered marginal in adverse climatic conditions. Sites on Dartmoor lying over 300m certainly prospered in periods of climatic optimum but the cooler and wetter summers of periods of climatic deterioration, their effect exaggerated by altitude and by the normally heavy rainfall of the area, would have led to the increased failure of both corn and hay crops here. On the clay-lands, merely some 20m above sea level, Goltho and its surrounding settlements were affected by the heavy rains and by the lack of evaporation which made the land unworkable owing to the exceptionally difficult soils of the area and their tendency to waterlogging and compaction. Although it is generally agreed that much land which could be considered marginal by existing standards of agrarian technology was brought into cultivation in the 12th and 13th centuries, and that much of this was the first to be abandoned after the Black Death, such an argument cannot be applied in the discussions on the desertions of Houndtor and Goltho. Both sites developed and prospered during the years of climatic optimum from the date of their foundation, in the 9th century, to the time the climatic deterioration occurred in the 13th century. Even if these settlements did become over-populated in the 13th and early 14th centuries, they should have been well able to have supported after the Black Death no smaller a community than that of circa 1000 to 1200, had the climate remained unchanged.

REFERENCES

Beresford, G. (1975) *The Medieval Clay-Land Village: Excavations at Goltho and Barton Blount.* The Society for Medieval Archaeology, Monograph Series No. 6, London.

Beresford, G. (1977) The excavations of the deserted medieval village of Goltho, Lincolnshire. *Chateau Gaillard: Etudes de Castellogie Mediềvale*, VIII, Actes du Colloque International tenu à Bad Muenstereifel 1976, 47–68.

Beresford, G. (1979) Three deserted medieval settlements on Dartmoor: a report on the late E. Marie Minter's excavations. *Medieval Archaeology*, 23, 142–6.

Beresford, M. and Hurst, J. G. (1971) *Deserted Medieval Villages.* Lutterworth Press, London.

Boswell, J. (1807) *The Journal of a Tour to the Hebrides with Samuel Johnson, LLD.* London, 4th Edition. See also S. Johnson (1825) *A Tour to the Western Islands.* London.

Bruce-Mitford, R. L. S. (1956) A Dark Age settlement at Mawgan Porth, Cornwall. In R. L. S. Bruce-Mitford (ed.) *Recent Archaeological Excavations in Britain*, Routledge, London.

Clarke, R. R. (1937) The Hebrides: A cultural backwater. *Antiquity*, XII, 261–189.

Denford, G. T. (1975) Economy and location of Bronze Age settlements on Dartmoor. *Bulletin of the Institute of Archaeology*, XII, 175–96.

Greig, O. and Rankine, W. M. (1956) A Stone Age settlement system near East Week, Dartmoor: Mesolithic and post-Mesolithic industries. *Proceedings of the Devon Archaeological and Exploration Society*, V, 8–25.

Grimm, P. (1939) Hohenrode, ein mittelalterliche Siedlung im Sudharz. *Veroffentlichungen der Landesanstalt fur Volkheitskunde zu Halle*, 11, 1–56.

Hamilton, J. R. C. (1956) *Excavations at Jarlshof, Shetland.* H.M.S.O., Edinburgh.

Lamb, H. H. (1972–7) *Climate: present, past and future*, Vol. I *Fundamentals and Climate Now* (1972), Vol. II *Climatic History and the Future* (1977). Methuen, London.

Lamb, H. H. (1978) Foreward. In M. L. Parry, *Climatic Change, Agriculture and Settlement.* Dawson, Folkestone.

Linehan, C. D. (1966) Deserted sites and rabbit-warrens on Dartmoor. *Medieval Archaeology*, X, 113-144.

Parry, M. L. (1978) *Climatic Change, Agriculture and Settlement*. Dawson, Folkestone.

Radford, C. A. R. (1952) Prehistoric settlements on Dartmoor and the Cornish Moors. *Proceedings of the Prehistoric Society*, XIII, 55-84.

Roussell, A. (1934) *Norse Building Customs in the Scottish Isles*. Williams and Norgate, London.

Staines, S. (1979) Environmental change on Dartmoor. *Devon Archaeological Society, XXXVII, Jubilee Conference Proceedings: Prehistoric Dartmoor in its Context*, 21-47.

CLIMATE AND SETTLEMENT IN LATE-MEDIEVAL CENTRAL ENGLAND

G. H. Dury
Emeritus Professor of Geography and of Geology/Geophysics
University of Wisconsin-Madison

Step-functional changes of climate, revealed by the analysis of instrumental records, is also indicated by British tree-ring series and by the documentation of the 1310s. Enquiry is needed into the possible effects of sudden change on villages which, eventually to be deserted, were significantly poor in 1334 and, in some instances, before Domesday.

Specialists on deserted medieval villages tend, as a group, to discount climatic change as a cause of desertion. Some historians appear to resent the intrusion of non-historians into historical studies. Paleoclimatologists, by contrast, are concerned in considerable part with climatic change; they can scarcely be blamed for wondering if some of the most drastic changes of, say, the last five thousand years may not have produced impacts upon human affairs. Particularly is this so, because many disciplines have in recent years become involved with environmental hazards, the relevant studies including, among other matters, the physical, financial, and social effects of natural disasters, and also human responses. In the background lie the resurgence of neocatastrophism, for instance in paleontology and sedimentology, and the wide discussion by social geographers and others of catastrophe theory. Two extreme viewpoints seem to exist: that of the committed opponents of determinism in all its forms, who regard changes in the physical environment as irrelevant to human affairs, or as of minor significance at the most, and who can quite properly demand to be told how physical changes actually work on human societies; and the viewpoint of those who find it incredible, or at least highly improbable, that natural changes in the environment should not have made their effects felt.

Needless to say, we shall make no progress if we start from a position of *parti pris*. The present paper is intended to offer for consideration two sets of propositions, each set independently susceptible of testing, and to suggest that, taken together, the propositions indicate that an enquiry on probabilistic lines might prove fruitful. The propositions are as follows:-

1) Short-term climatic change is typically frequent and step-functional;
2) Many of the villages eventually to be deserted were significantly poor and/or below average in population numbers,

prior to the Black Death; some of them were identified as poorly endowed with agricultural resources, even in pre-Conquest times.

The term *significantly* is here, as subsequently, employed in the statistical sense. The line of potential probabilistic enquiry is indicated by the statement that, on ecological grounds, one would expect the weakest members of a community to suffer the most severely in times of environmental stress. Weakness in the present context means, for a given village, limited extent of arable land, low ploughland quality, low material wealth generally, and presumably also low population total. The environmental stress to be discussed is that of sudden climatic shift: this statement is in no way meant to deny the existence, or the incidence, of additional stresses, including climatically-independent social and economic stresses. What is needed for the type of enquiry advocated, and what is as yet unavailable, is a parallel listing of the incidents of village desertion and of the timing, direction, and amplitude of changes in climate. Until such a listing can be drawn up, we cannot go further than to state an hypothesis - for example, the null hypothesis that climatic shift had no significant effect on village desertion. We can, however, scrutinise the evidence for the already-stated propositions. An obvious difficulty is that the evidence for climatic change consists very largely of surrogate information, in respect of the period preceding continuous instrumental observations. A problem quite unforeseen when the present analysis began is that climatic data, whether surrogate or instrumental, indicate that significant changes, for instance in annual regimes, occur with surprising frequency - say, on the average of ten to twenty years. For purposes of comparison with climatic reconstructions, the mesh of historical information would need to be very fine indeed.

COMPARATIVE STATUS OF VILLAGES EVENTUALLY TO BE DESERTED

My statistical analysis so far extends through a group of Midland and near-Midland counties: Bedfordshire, Buckinghamshire, Cambridgeshire, Derbyshire, Gloucestershire, Huntingdonshire, Northamptonshire, Nottinghamshire, Oxfordshire, Warwickshire, and the East Riding of Yorkshire, as these existed before the boundary revisions of the 1970s. The original intention was to investigate an area that could be broadly identified with the Central England for which Lamb (1966) has reconstructed a series of temperature values. The initial (and, as it turned out, misguided) research design included the analysis of place names: hence the omission of Leicestershire, for which no place-name volume had appeared, and the inclusion of Gloucestershire and the East Riding by way of extending the original data base.

Desertion rates for the area concerned, reckoned perforce from such lists as Domesday and the *Nomina Villarum*, appear to have amounted to about 2% of all villages in the period between the mid-11th century and

TABLE 1

1334 Lay Subsidy: Mean Levies for study area in pounds

County	TBDs	Comb TBD & NTBD	Comb NTBD	All NTBD	Singles NTBD
BEDS	6• 2.62	12• 2.23	12• 3.01	65• 4.96	53• 5.40
BUCKS	20• 2.22	26• 1.51	48• 2.03	123• 2.69	75• 3.08
CAMBS	5• 5.26	3• 3.19	13• 2.77	56• 4.37	43• 6.15
DERBYS	23• 1.06	1• 1.10	22• 0.92	173• 1.66	151• 1.77
GLOS	41• 1.58	19• 2.09	42• 2.61	246• 2.65	204• 2.66
HUNTS	10• 3.76	4• 2.12	9• 2.88	52• 4.82	43• 5.23
NORTHANTS	17• 2.04	34• 2.10	45• 2.33	209• 2.53	164• 3.56
NOTTS	27• 1.88	28• 1.11	40• 1.71	219• 2.11	179• 2.20
OXON	59• 2.22	30• 2.17	35• 2.56	228• 3.67	193• 3.87
WARKS	62• 1.31	36• 1.72	41• 1.56	232• 2.35	191• 2.53
YORKS ER	75• 1.40	45• 1.00	37• 1.40	247• 2.45	210• 2.58

notes: • indicates number of settlements; TBD analysis includes combined all TBD in addition to TBD singles; Comb NTBD excludes TBD/NTBD combinations; All NTBD is the weighted mean of the two values on either side.

the mid-14th. The Lay Subsidy Rolls of 1334 (Glasscock, 1975) supply a *terminus ad quem*. The indicated period is generally regarded as one of improving (alternatively, ameliorated) climate and of increasing population. The Lay Subsidy data will be used to assess the comparative status of villages eventually to be deserted (T.B.D.s) and those not known to have undergone desertion (N.T.B.D.s). Whereas the Domesday record, containing only centres of administration, may have omitted perhaps one-third of the settlements that actually existed in 1086, the Lay Subsidy rolls appear to be less exclusive. Analysis for the indicated counties has been limited to those hundreds in which known T.B.D.s occur. For those hundreds, places listed merely as *cum membris* or *cum soca* have had to be left out of the account, the members or sokes being unnamed. Numbers of T.B.D.s are doubtless included within these anonymous places, but even with their exclusion, about 75% of known T.B.D.s have been identified in the Subsidy listings.

The general observation that T.B.D.s were poor and small at the time of desertion has been frequently made. The matter is discussed for instance by Beresford (1971) who, in the strict context of their morbid pathology, comments that there is every evidence that vulnerable villages were small in the early fourteenth century. It seems as desirable, however, as it is extremely simple, to go beyond the mere comparison of raw averages. Difference-of-means tests serve well in the analysis of the 1334 data, as they also do in the analysis of similar data for other periods and in the analyses, to be presented later, of climatic time series.

For purposes of difference-of-means testing, the 1334 data, selected as indicated above, have been grouped into T.B.D.s and N.T.B.D.s. The combinations originally made for tax purposes demand an elaboration of the grouping to allow for the combination of, say, a T.B.D. with an N.T.B.D., or of two N.T.B.D.s as a pair (Tables 1 and 2). The cutoff

TABLE 2

Statistical Comparison of 1334 Levies: Values in Pounds

	Singles			Pairs		
BEDS	TBD	2• 3.32	0.010>P	WTBD	8• 2.34	0.025>P
	NTBD	53• 5.40		NTBD	6• 3.01	
BUCKS	TBD	20• 2.22	0.010>P	WTBD	13• 1.51	0.050>P
	NTBD	75• 3.08		NTBD	21• 1.98	
CAMBS	TBD	5• 5.26	0.050>P	WTBD	0•	
	NTBD	43• 6.15		NTBD	n/a	
DERBYS	TBD	23• 1.06	0.001>P	WTBD	1• 1.10	not
	NTBD	151• 1.77		NTBD	11• 0.92	worked
GLOS	TBD	37• 1.61	0.001>P	WTBD	6• 0.97	0.001>P
	NTBD	204• 2.66		NTBD	13• 2.28	
HUNTS	TBD	8• 3.08	0.010>P	WTBD	3• 3.57	P>0.700
	NTBD	43• 5.23		NTBD	3• 2.87	
NORTHANTS	TBD	11• 2.25	0.005>P	WTBD	19• 2.01	P>0.050
	NTBD	119• 3.56		NTBD	21• 2.41	
NOTTS	TBD	27• 1.88	0.250>P	WTBD	8• 1.03	0.020>P
	NTBD	179• 2.20		NTBD	14• 1.73	
OXON	TBD	55• 2.26	0.001>P	WTBD	13• 2.29	P>0.500
	NTBD	193• 3.87		NTBD	16• 2.51	
WARKS	TBD	52• 1.49	0.001>P	WTBD	11• 1.29	P>0.800
	NTBD	191• 2.53		NTBD	12• 1.63	
YORKS ER	TBD	61• 1.58	0.001>P	WTBD	16• 1.67	P>0.500
	NTBD	210• 2.58		NTBD	12• 1.46	

WTBD = with one or two TBD

level applied in testing was P = 0.05. For the area as a whole, the levy on T.B.D.s averaged 62% of the levy on T.B.D.s, the difference being statistically significant. Paired T.B.D.s, and pairs composed of one T.B.D. and one N.T.B.D., were assessed significantly lower than were paired N.T.B.D.s. Furthermore, the T.B.D.s were significantly more likely to be combined, for purposes of the Subsidy assessment, than were N.T.B.D.s. The county breakdowns suggest possible regional variation, which is still under study. It is already apparent that single T.B.D.s were assessed significantly less than single N.T.B.D.s, except in Nottinghamshire. The difference of mean levy between paired T.B.D.s (lower) and paired N.T.B.D.s (higher) is significant within counties only for Bedfordshire, Buckinghamshire, Gloucestershire, and Nottinghamshire. I owe to an anonymous reviewer the reminder that the 1334 Subsidy was a lump sum negotiated for each vill, and that a main variable in determining the size of the lump sum was the number of inhabitants. Significantly low assessments may therefore be read to indicate a combination of communal poverty with low communal population, the latter going at least part of the way to explain the former.

Scope for similar analysis of Domesday data is limited by the already-mentioned clumping of settlements in the 1086 returns and by the complexities attending any attempt to use hidation statistics. These latter are notably affected by measures of tax relief, and quite possibly also by taxation rackets. Nevertheless, something can be tried with per-hide values and per-ploughland values. Analyses of these for the study area is not yet complete. In Huntingdon and Berkshire, although not in Bedfordshire, Buckinghamshire, Cambridgeshire, and Oxfordshire, the data given in Domesday for T.R.E. values (in the time of King Edward: we may understand 1065) show T.B.D.s as averaging significantly less than N.T.B.D.s, both in per-hide and in per-ploughland ratings. The T.B.D./N.T.B.D. ratios are 100 to 75 or 80. For Huntingdonshire and Berkshire, then, the original rough-cut assessment, which may have gone back to the first general Danegeld of the 10th century, seems to have overestimated the farmland potential of the T.B.D.s, while the revised assessments of the mid-11th century rated that potential as significantly low. To this limited extent - limited, that is to say, to two of the six counties analysed so far - there seems to be evidence that some T.B.D.s had got off to a bad start in the first place.

Far more work is obviously required on the site characteristics of individual T.B.D.s. To mention only one possible line: a group of T.B.D.s on the Warwickshire/Northamptonshire border is located on the Lower Lias outcrop, which, devoid of subsurface water, is wholly dependent on rainfall. Its soils are bentonitic, fearsomely sticky in the wet and forbiddingly hard in the dry. To judge from the incidence of livestock bloat, they are deficient in one or more trace elements. A molybdenum deficiency, to name one for investigation, is conceivably carcinogenic, both for humans and for livestock, since it means reduced levels of vitamin C in plants and enhanced nitrite levels in animal tissue. In this and related contexts, however, we should be investigating potential causes of poverty and low population - that is, of the vulnerability which appears to have caused desertion to be selective. In what now follows, I shall discuss evidence for sharp and frequent climatic change, intending eventually to suggest that such change may have selectively harmed villages that were already small and poor.

CLIMATIC DETERIORATION AND STEP-FUNCTIONAL CHANGE

Lamb's reconstructions (Lamb, 1966) include a cresting of temperature values in Central England in about 1200-1250, a sharp decline to about 1400, and a gentler decline thereafter to about 1600. Reference to climatic data for the East Midland area (Dury, 1963) indicates for 1250-1350 a reduction of about one week in the growing season, with a decrease of 200 degree-days above a base of 4.4°C. Corresponding values for 1250-1500 are three weeks and 300 degree-days.

Parry (1975), tracing the fall of the cultivation limit during the same general interval on the Lammermuir Hills, finds the limit of cereal farming descending through 50m of altitude between 1250 and 1350, and through 200m by the 1600s. His orographic findings are readily brought to agree with Lamb's reconstructions of temperature change, with the aid of reasonable assumptions about lapse rates. He has also usefully considered the likely shifts of the end-of-summer moisture surplus; but it seems fair to suggest that the principal factor considered is temperature, just as it has been in discussions of the well known disappearance of vineyards from the English Plain. Numerical manipulations of reconstructed precipitation data, at least those data which apply to annual regimes and to high summer and deep winter, seem to produce inconclusive results. Thus, when Lamb's reconstructed values of annual and extreme-season rainfall are used in calculations based on the equations of Thornthwaite, the exercise fails to retrodict any significant increase in surface moisture, such as may have set in about 1250.

Hurst (1971) has called attention to the changing design of the foundations of peasant houses during the later 13th and the 14th centuries. In the stone belts of the English Plain, basal timbers came to be laid on stone footings instead of being set directly on the ground, as previously. In the clay belts, houses came to be built on earthen platforms made with the spoil from larger drainage ditches. These ditches, replacing the shallow gutters of earlier years, themselves needed to be cleared out from time to time. Although any interpreter is at liberty to appeal to the evolution of building design, Hurst himself inclines to the view that climatic change was involved at least to some extent.

One particular, and sudden, increase in surface wetness is well documented. This was the well known, and often described, agricultural calamity of the second decade of the 14th century. It would be tedious to retail the documentation of a firmly established set of climatic, agronomic, and social events. Repeated crop failures, for example, in Germany and in southern England, were accompanied by famine and even by cannibalism. As has happened on other occasions, such as with the termination of the Sahel droughts and of the extensive flooding in recent years of the Australian Centre, the sudden change in the regime of seasonal weather set in with great initial violence. It can readily be shown that the change was step-functional.

The Account Rolls of the Bishopric of Winchester (Titow, 1960), which extend from 1219 to 1349, supply usable weather data on 83 of the 121 years, for some or all of the three farming seasons of summer, autumn, and winter. Reference to analyses of twentieth-century instrumental data suggests a test for breakdown into blocks of 50 years, or fewer, wherein climate differed in some respect, or bundles of respects, from that of preceding and succeeding blocks. For purposes of analysis, graphing and computer operation serve equally well. The intervals 1219-45, 1246-1309, and 1310-49 separate themselves out.

TABLE 3

Seasonal Weather in Manors of the Bishopric of Winchester

Year	Win (Dry)	Sum (Dry)	Aut (Dry)	Win (Wet)	Sum (Wet)	Aut (Wet)
1219						W
1221		d				
1224						w
1225	h	D				w
1226	h					
1227						w
1228		D				
1233						W
1236	h	D				
1245		d				w
1246					w	w
1247		D	d			
1248	h	D				
1252	½h	D				w
1253		d				
1254	h				½W	W
1255	h					
1257	h					
1258					w	w
1263		D				
1266		d				
1268	h	d				w
1269	h	d				w
1271	h			w		
1272	h	D	d			
1273				w	W	
1274		D				
1275		½d			½w	w
1278		D				
1284		D		w		
1285	h	D				
1286		d				
1287		D				
1288		D				
1289		D		w		w
1290		D		w		w
1291		D				W
1292	H				w	
1293		d				
1297		D				
1298		½D			½w	

Year	Win (Dry)	Sum (Dry)	Aut (Dry)	Win (Wet)	Sum (Wet)	Aut (Wet)
1300		D				
1302		½D				½w
1303		d				
1305		D				
1306		D				
1308						W
1309		D				
1310		d		w		
1311		d				
1313				w		W
1314	h					W
1315				w	W	W
1316				w		w
1317				w		
1318		D				
1319				w	w	W
1320						W
1321		d				
1325		D		w		
1326	h	D				
1327	h	D				
1328				w		
1329		D				
1330	h				w	W
1331		D		½W		
1332		d				
1333		D				
1335	½d					
1336		D		w		
1337				½w	w	
1338		d		w		W
1339	h	d		w		
1340	h	d				
1341	d					
1342		d				w
1343		D				
1344		D		w		
1345				w		W
1346		d		w		
1347		d		w		W
1348				w		w
1349				w	w	w

Their percentages of reportedly severe winters are, respectively, 30, 27, and 17; of reportedly wet winters 0, 21, and 53; and of wet autumns, 60 (in actuality, 6 out of only 10 reported), 29, and 40. Of the reported autumns for 1310-49, most were either very wet, or wet and long.

If reported information is scored, then the distinctive characteristics of the indicated intervals can be confirmed by the technique of plotting cumulative deviations from the mean and difference-of-means tests can be applied. The reported sequence is listed in Table 3. For each season, the attributes have been scored individually on a scale of 2 to zero. For example, for summer wetness, no report (for a year with extant records) = 0; part season wet = 0.5; whole season wet, or half season very wet, = 1.0; whole season very wet = 2.0. Difference-of-means and difference-of-variance tests have been run on the mean scores for the indicated intervals, the cutoff P = 0.05 being applied throughout. From 1219-45 to 1246-1309, and from 1246-1309 to 1310-49, significant increases occurred in the frequency of wet winters and in the frequency of wet summers (or infrequency of dry summers). From 1246-1309 to 1310-1349, both the frequency of wet seasons in total and the variance of seasonal wetness increased significantly.

Insofar as the Winchester records are qualitative, they are probably to be regarded, at least to some extent, as surrogate in nature. Tree-ring data as climatic indicators are entirely surrogate in nature, with formidable difficulties attending their interpretation (Fritts, 1976). Site complications apart, however, it seems unlikely that many would deny that the main controls on rates of tree growth, especially when this growth is assessed on a regional basis, are climatic/ meteorological. To work back from the tree ring record to a reconstruction of climate is altogether a different matter. In any but the easiest sites, as where growth is controlled by a single variable such as moisture, it calls for multivariate analysis. Moreover, trees possess memories in the same way that computers can be said to possess memories: they store incoming information for later use. Growth in seasonal climates is not merely a matter of the climate experienced in a particular growing season. Some at least of the West European tree ring series appear quite insensitive to the hydrometeorological events of the 1310s. One might conclude, provisionally, that such temporal changes in growth rates as can actually be demonstrated relate to changes in energy inputs. This whole matter, however, is beside the immediate point, which is that rapid changes in growth rates of west European trees can be readily demonstrated. If the changes were climatically provoked, then it follows that climate also underwent rapid change.

The records published by Baillie (1977a, 1977b, 1977c) for northeast and east-central Ireland and for south-central Scotland, have been tested for step-functional variation. The result, in each case, is the breaking down of the series into a sequence of square waves (Fig. 1). An horizontal segment, or tread, indicates a mean value about which, for a certain run of years, observed values fluctuated. Such a run is herein referred to as a block. Risers (and fallers) between one block and the next mark step-functional change between the respective means of successive blocks.

The analytical technique employed depends, to begin with, on the determination of cumulative departures from the mean of an entire series. Since Baillie's data express ring width by index values (wherein the mean for an entire series = 100) the series have already been converted to the stationary form. Calculation carries an inbuilt check of accuracy since the final cumulative error must be zero. When cumulative departures are graphed, the graph typically displays rising and falling limbs with a few more or less horizontal limbs interspersed here and there. A rising limb indicates that, during the particular part of the series, observed values were running above the series mean;

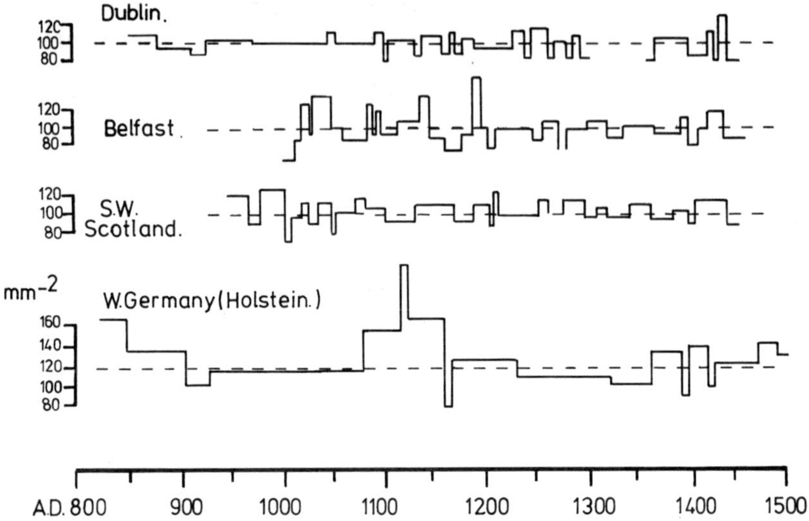

Figure 1 - Results of the step-functional analysis of oak tree-ring series.

a falling limb indicates a run of values generally below the series mean; and a horizontal limb relates to values averaging close to the series mean.

Breaks between segments of different slope are characteristically abrupt. For the block of years corresponding to a given segment, linear regression of cumulative departures against serial positions tests the proposition that the segment may fairly be represented by a straight line. Typical r values turn out to be ±0.975 or better, with P = 0.05 or better, except for near-horizontal segments. The block mean for one of these segments has been accepted as significantly different from the series mean, only where r is significantly different from zero. Elsewhere, the $\bar{y}$ values of the regression equations indicate where the treads should be drawn. A break between one block and the next has been recognised, only where the respective block means differ at the 0.05 level of probability, or better.

The results indicate a gross lack of correspondence in detail between any pair drawn from the Baillie series and between any member of these series and Hollstein's series for West Germany. This circumstance promises poorly for any general reconstruction of climatic shifts in the West European area by means of dendrology. The point immediately at issue, however, is neither that of causes of variation in growth rate nor that of correlation from series to series. It is the frequency of change in the climatic complex which must surely be regarded as

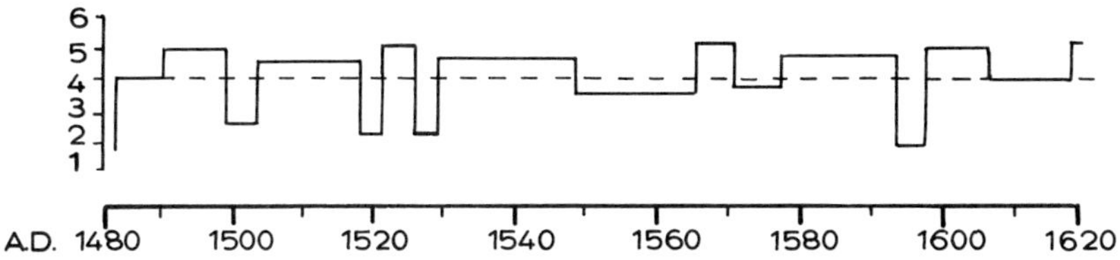

Figure 2 - Results of step-functional analysis of the Hoskins data on grain harvests.

influencing growth rate. For the Belfast and Scottish series between 1200 and 1450, the mean length of block is some 20 years. The climatic complex could apparently change from one generation to the next.

Although the data of Hoskins (1953) relate to a period subsequent to that when most village desertions occurred, they may usefully be made to supply another example of step-functional change. Hoskins refers data of wheat prices for 1480-1619 to the 31-year moving average, arriving at an assessment of harvests that ranges from abundance to dearth; a good harvest, for instance, is indicated by prices between 10% and 30% below the moving average. When his assessments are scored, with abundance = 6 to dearth = 1, and when the sequence is analysed for cumulative departures from the mean, step functions again appear (Fig. 2).

Needless to recall, the relationship between yield and price can be complexly, and endlessly, argued. But if any relationship of any importance be conceded, then the step-functional sequence in Figure 2 indicates a step-functional variation in yield. As Hoskins says, between one-half and two-thirds of the population, during the period considered, were wage earners and a considerable proportion of the remainder were subsistence farmers. Some 80 to 90 per cent of cash wages was spent on food and drink. About one-third of the population was on, or near, the poverty line: another third was below. Harvest was the fundamental fact of economic life.

Now, the mean length of block in Figure 2 is some 8 years. As is usual in analyses of this kind, block means run for the most part distinctly above or distinctly below the long-term mean. The general probability seems to have been that during a given run of years, growing and harvesting seasons would be mainly favourable to highly favourable, or mainly unfavourable to really poor. To any reader who might wish to protest that we are really discussing the state of the market, as opposed to harvest characteristics, one could perhaps suggest that a challenge should be directed at Hoskins's original interpretation rather than at the treatment of his findings which is presented here. In any event, abrupt market fluctuations at short intervals, during the period with which he deals, can scarcely have failed to affect grain producers and to have affected the most severely those with the least reserves. We may guess that, in earlier periods, the importance of the market was reduced and that of yield enhanced. The analysis of the surrogate

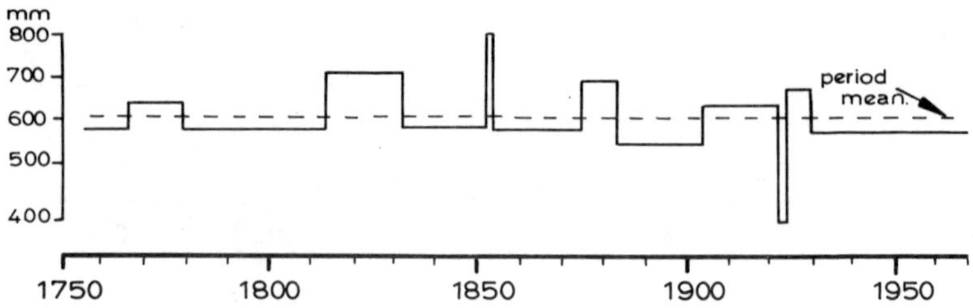

Figure 3 - Results of step-functional analysis of precipitation data for London.

information of rates of tree growth has already indicated, for those same earlier periods, abrupt changes in the total effects of climate.

The pattern of variation obtained by the analysis of surrogate data is identical with that obtained by the analysis of instrumental data. Figure 3 presents the results of analysis of annual precipitation at London from 1750 onwards. Although the use of transforms (e.g. cube-root values) can be preferable in the analysis of precipitation data for dry climates, the raw data for London respond well to the determination and graphing of cumulative deviations, linear regression within blocks of observed departures on serial position, and difference-of-means tests between blocks. While there is no significant difference of variance between blocks (at P = 0.05), means between successive blocks do differ significantly. The type of variation through time displayed by the London precipitation data is precisely that identified for tree rings and grain harvests.

The block length for the London precipitation series averages about 16 years. Analysis of monthly and seasonal data could be expected to reveal breaks additional to those which affect an annual series. The average block lengths for precipitation at Sydney, N.S.W., for instance, are somewhat over 8 years for the annual series and close on 11 years for the individual monthly series; but the incidence of change in these individual series means that, in the total record, some monthly regime or another has changed, at an average interval of about one year (Dury, 1980). The frequency of change of regime for whole seasons remains a matter for future study.

CONSIDERATIONS OF RANDOMNESS

Discussing the distribution of deserted villages in the lowlands, Beresford (op. cit. 1971) concludes that some important selective factor must have been at work which, being still undiscerned, may conveniently be labelled random. A systems analyst might think it inappropriate to label as random the processes operating within some cascade or black box. Randomness can however be brought to bear, in more than one way, on the probability of village extinction.

Curry (1962) discussed climatic change as a random series, dealing among other matters with the question (and probability) of persistence. Cumulative random departures from a given mean can be shown to simulate, in many instances, the corresponding departures in real series. Considerable portions of the square-wave sequences drawn for the Baillie data on tree rings can, for example, be matched from the results of analysis of random data. The distribution of block lengths in the tree-ring and climatic series matches the distribution obtained from the analysis of random series. Although sudden step-functional changes in midlatitude climate can be investigated in relation to abrupt changes in the pattern and location of the jet streams (Lacher, McArthur, and Buznya, 1977), there seems to be no *a priori* reason to search for a wholly deterministic model, as opposed to a probabilistic model, of jet-stream behaviour. For a number of purposes, the individual travelling pressure systems of the Westerlies belts are regarded as random perturbations. Analysis of residuals from step-functional precipitation models indicates normal - i.e. random - variation about block means. That is to say, variation both within and between blocks can be simulated by random techniques.

Square waves may or may not be superimposed on progressive trends. Cycles will not be discussed, because the smoothing of data on which cyclic analysis often depends can itself be relied on to generate sine waves; and such waves, as can readily be demonstrated, can themselves be simulated by random means. So, for that matter, can progressive trends; but the emphasis in this paper is placed, as will have been clear, on abrupt changes from one short-term regime to another. Even with progressive trends, the application of catastrophe theory would lead to the prediction of step-functional response. The square-wave sequences inferred from real data and from surrogate records suggest that many medieval villages, already close to the threshold of non-viability, may have been far more seriously exposed to climatic hazard than is often realised.

Randomness can also be considered in relation to population dynamics. For a considerable interval from the mid-14th century onward, the population total of England is thought to have changed little. Its state was low-fluctuating. Very little net change occurred; any net increase that can be read from the available records can be absorbed in the estimated increase of the population of London. For rural areas, the condition was one of steady state. It is however scarcely possible to imagine that no villages failed to experience net growth, say between 1350 and 1450. But if some grew, then others must have declined. A random upward drift of population total in some villages can be imagined as offset by a random downward drift in others. A possibility of extinction is built into the random-walk model of village populations. The question, put from time to time, of where the people of deserted villages went to, may find a partial answer in the suggestion that they simply died off.

An obvious related possibility is that genetic drift may have contributed to the decline of some villages. This possibility comes inevitably to mind in the general context of population dynamics. Isolation and inbreeding were certainly producing undesirable genetic effects in parts of Lincolnshire as late as the 1930s, while analyses of parish registers are available for parts of the rural Midlands to show that, during the mid-19th century, marriage partners typically came from a home village or from a next-neighbouring village. A closed community is more liable to genetic drift than is a community with free external exchange; and one may guess that the probability of unfavourable drift is greater than the probability of favourable drift. It might however be misleading to extrapolate back from 1900 to 1400; one of my reviewers informs me that exogeny in medieval villages is suggested by recent work to have been far more common than might, on general grounds, be expected, and that turnover rates, e.g. of 75% in 25 years, may have run higher for T.B.D.s than for N.T.B.D.s. We appear, therefore, to be referred back to the working of a low-fluctuating population model.

CONCLUSION

The general mechanism of any linkage between environmental change and demographic response remains to be determined. No determination is attempted in the present paper. It has been shown, however, that climatic time series can be analysed into step-functional models and that similar models apply to surrogate data, including rates of tree growth and assessments of harvests. Quite apart from secular trends, with durations measured perhaps in centuries, climatic series vary step-functionally in the short term, with regimes changing at average intervals of some 10 to 20 years according to whether annual, seasonal, or monthly series are considered. Step-functional change of regime, progressive variation, and cyclic fluctuation can all be simulated by random means.

For at least a number of lowland counties, the villages that would eventually be deserted were, in 1334, significantly smaller/poorer, as a group, than were villages for which no desertions are known. In two counties, the eventually deserted villages were significantly poor in agricultural resources and/or productivity and/or potential as early as the mid-11th century.

The steady-state model of low-fluctuating population total can accommodate both the random upward drift of some individual village populations and the random downward drift of others. On general grounds, one would expect the weakest members of a village system to be the most liable to suffer from environmental stress, the abrupt onset of which is indicated by the analysis of climatic and surrogate data.

ACKNOWLEDGEMENTS

I gladly acknowledge funding by the Graduate School of the University of Wisconsin-Madison (Project No. 181441), which supported conference participation, computing services, and drafting. R. J. O'Brien, Project Assistant, cheerfully undertook a great deal of laborious extraction of basic data and was also heavily engaged in some of the regression analysis. Professors Waltraud Brinkman (U.W.-Madison), Grace Wahba (U.W.-Madison), and Barbara Bell (Harvard) have been generous with counsel on the treatment of time-series data: I owe to the last of these the advice to consider the use of mass residuals, with which portions of her own work deal. I am also grateful to certain reviewers. Two of my debts in this connection are acknowledged in the text; a third is owing to an anonymous historian who, reviewing a related manuscript, directed my attention to recent work that greatly illuminates the problems involved in the use of place names.

REFERENCES

Baillie, M. G. L. (1977a) The Belfast oak chronology to A.D. 1001. *Tree-Ring Bulletin*, 37, 1-13.

Baillie, M. G. L. (1977b) Dublin medieval dendrochronology. *Tree-Ring Bulletin*, 37, 13-20.

Baillie, M. G. L. (1977c) An oak chronology for south-central Scotland. *Tree-Ring Bulletin*, 37, 33-44.

Beresford, M. (1971) A review of historical research (to 1968). In M. Beresford and J. C. Hurst (eds.) *Deserted Medieval Villages*, Lutterworth Press, Guildford and London.

Curry, L. (1962) Climatic change as a random series. *Annals, Association of American Geographers*, 52, 21-31.

Dury, G. H. (1963) *The East Midlands and the Peak*, Nelson, Edinburgh.

Dury, G. H. (1980) Step-functional changes in precipitation at Sydney, N.S.W., *Australian Geographical Studies*, 18, 62-78.

Fritts, H. C. (1976) *Tree Rings and Climate*, Academic Press, London, New York and San Francisco.

Glasscock, R. E. (1975) *The Lay Subsidy of 1334*. Oxford University Press, London.

Hoskins, W. G. (1943) Harvest fluctuations and English economic history 1480-1619. *Agricultural History Review*, 2, 28-46.

Hollstein, E. (1965) Jahrringchronologische Datierung von Eichenholzern ohne Wald Kante (Westdeutsche Eichenchronologie) *Bonner Jarhbucher* 165, 1-27.

Hurst, J. G. (1971) A review of archaeological research (to 1968) In M. Beresford and J. G. Hurst (eds.) *Deserted Medieval Villages*, Lutterworth Press, Guildford and London.

Lacher, R. G., McArthur, R., and Buznya, G. (1977) Catastrophic changes in circulation flow patterns. *American Scientist*, 65, 614-621.

Lamb, H. H. (1966) *The Changing Climate*. Methuen, London.

Parry, M. L. (1975) Secular climatic change and marginal agriculture. *Transactions of the Institute of British Geographers*, 64, 1-13.

Titow, J. (1960) Evidence of weather in the account rolls of the bishopric of Winchester, 1219-1350. *Economic History Review*, 2nd series, 12, 360-407.

CLIMATE AND ECONOMY IN EIGHTEENTH CENTURY ICELAND

A. E. J. Ogilvie
Climatic Research Unit
University of East Anglia

Unpublished 18th century reports sent annually by officials in Iceland to the central government in Copenhagen give details of trade, the state of the hay-harvest, fishing catch, and weather and climate. These reports have been used in an analysis of the Icelandic climate and economy of the 18th century. Their evidence suggests that the hardship experienced in Iceland at this time was largely due both to the restrictive trading monopoly enforced by the Danes and the adverse climate of the period.

The eighteenth century in Iceland was a time of extreme hardship. This was undoubtedly the result of both economic and environmental factors. In this discussion, however, the influence of the climate will be emphasised and some effects of short-term variations in climate (on a time-scale of a decade or less) on certain aspects of Icelandic society, especially activities such as farming and fishing, will be considered.

The impact of climate on society varies enormously from place to place and time to time. It seems safe, however, to assume that studies of the impact of climate will be most successful either where climatic "extremes" are likely to be experienced or in areas which are marginal for settlement and agriculture (Parry, 1978). Both these criteria are applicable to Iceland and these factors combine to make Iceland an ideal choice for this type of study.

CLIMATE AND ENVIRONMENT OF ICELAND

In spite of being in a maritime situation (maritime regions generally have more moderate and less variable climates than continental regions) the climate of Iceland varies considerably on all time-scales. This is due to Iceland's geographical location. Situated just south of the Arctic circle and the cold Polar regions Iceland is subjected alternately to influxes of Polar air from the North and warmer air from the Atlantic. Iceland is also close to one of the main centres of cyclonic activity in the Northern Hemisphere (the "Iceland Low"). In addition, Iceland is affected by two contrasting ocean currents; the warm Irminger current which flows clockwise around the south and west coasts, and the East Greenland polar current which flows southwards round the east coast. The warm ocean currents serve to ameliorate Iceland's climate and hence facilitate human settlement.

Iceland lies close to the seasonal boundary of the Arctic drift ice which is carried southward by the East Greenland current. The extent of the ice varies considerably from year to year and century to century; in the 20th century it has been comparatively rare, in the 18th century it was a frequent visitor. The ice most frequently affects the northwest, north or east coasts; it is rare for it to reach the southwest or south. The presence or absence of the sea ice has a considerable influence on the climate of Iceland primarily because proximity of the ice means a fall in both land and sea temperatures, (Bergþórsson, 1969). Sea ice occurs most frequently off the coast of Iceland from late winter to early spring.

The drift ice has always been of great importance for the human population. In the past it was thought to bring certain benefits; for example, its presence was associated with an increase in the amount of driftwood reaching the shore (Iceland has always been short of timber) and, also, the large numbers of seals which sometimes accompanied the ice provided useful products. Whales, too, were sometimes left stranded by the ice, and from early settlement times there have been laws regarding claims to the carcasses. On the whole, however, the ice has had a negative, even a disastrous effect on human settlement in the past. When ice stays into the summer the length of the growing season is significantly reduced. The lowering of temperatures brought about by the ice has an adverse effect on the growth of vegetation, especially the vital grass-crop. If the ice comes close into the shore and particularly if it becomes land-fast it destroys useful shore vegetation and shellfish (Friðriksson, 1969, 149). The presence of sea ice also delays access to open water for fishing boats, and to harbours for foreign trading vessels.

Much of Iceland's environment is harsh and inhospitable and human settlement is only to be found in the more fertile coastal lowland and valley regions. The central region, which has always been uninhabited, consists of a 500-800m high plateau with numerous peaks and smaller plateaus 1200-1800m high rising above it. Glaciers account for approximately 11.5 per cent of the total area of the country. Of these the largest is Vatnajökull in the southeast which is about 8,400 square kilometres in area and 1000m thick at its centre (Iceland 874-1974, 8). Cultivatable land has always been limited and today only 25 per cent of the dry land mass has a vegetation cover at all. Communication has also been difficult due to poor roads and the large and numerous rivers. Such factors are exacerbated by difficult weather conditions. To compound all these difficulties Iceland is also greatly affected by volcanic activity.

DOCUMENTARY EVIDENCE

The study of the consequences of climatic and environmental changes in Iceland is possible because there are reliable historical sources both of climatic and socio-economic data available from an early period. By the 18th century, however, they become fuller and more numerous than in earlier years. The documentary evidence available from this period includes annals, diaries, accounts written by native and foreign writers, and private and official correspondence. Other types of data such as land registers, parish records, censuses and trade figures are also available. In the following analysis particular use is made of certain manuscript sources; official letters sent by Icelandic representatives to the central government in Copenhagen ("Bréf til Stiftamtmanns", abbreviated here to B.S.). Iceland had been under Danish rule since 1380 and officials were, from time to time, required to send in reports on the state of the country. By the early 18th century attempts were made to systematize this kind of information and it became common practice for the official or "sheriff" in each of Iceland's 22 jurisdictional districts to send an annual report to the Governor of Iceland in Copenhagen. These letters contain information on legal matters, problems with regard to trade, the state of the hay-harvest, fishing and information on weather and climate. The letters quoted here are taken from the north and west of Iceland. This is because an analysis of letters from the east and south was not yet complete. However, although regional differences were important, there is no doubt that the letters quoted reflect conditions which applied to the whole of Iceland.

POLITICAL AND ECONOMIC ASPECTS

Union with Denmark and the Trade Monopoly

Iceland became part of the Kingdom of Denmark in 1380, when Norway, Denmark and Sweden joined together in the Kalmar Union (Iceland had belonged to Norway since 1262). This political accident was to cause Iceland much difficulty, mainly because of the restrictive trade policies applied to Iceland by the Danish Kings who, from an early period, sought to acquire trading privileges in Iceland. In 1602 the Danes established a monopoly of trade with Iceland which lasted until 1787. The monopoly was held at different times by different groups or by individuals but it was never popular with the Icelanders and, on the whole, it had a detrimental effect on their economy. They were not self-sufficient and therefore relied heavily upon imports of various essential goods which they did not have themselves: such as flour, timber and metals. As long as the monopoly lasted there were numerous complaints from the Icelanders, about the poor quality of the goods the merchants brought; the lack of suitable goods; the high prices of the imported goods coupled with the low prices the Icelanders were offered for the goods they had to sell; or the merchants' poor treatment of the Icelanders. The merchants also had their complaints, saying, for example, that the Icelanders did not properly prepare the goods they exported. However, there seems little doubt that it was the Icelanders who had the greater cause for complaints: the Danes, having virtually no competitors for the Iceland trade could largely do as they pleased. This letter from sheriff Bjarni Halldórsson of Húnavatnssýsla, dated 1754, reflects the difficult situation:

> *"The only thing of importance that I have to recount concerns the past severe winter which has caused great misery to the inhabitants here in the district as they have lost almost all their livestock and means of subsistence and they were not able to buy flour and foodstuffs from the merchants, let alone obtain them on credit, the monopoly's control being as it is. This year the behaviour of the merchant has been even more absurd and contrary than in previous years; in this time of dearth they have let people travel to the trading-places 2 or 3 times. Sometimes they were turned away with scornful words, on other occasions made to wait there for no reason at all, as it suited the merchants, only to finally be refused unless they had goods to pay with at once. However, their woollen goods were rejected and they were rudely forced to give up the few remaining cattle they had to live off as payment."*
>
> (B. S. Húnavatnssýsla, 1754.)

Farming and Fishing

Until comparatively recent times the main occupation in Iceland was farming. This was based on animal husbandry as the climate generally proved unsuitable for growing grain crops. Sheep, cattle and horses were the livestock most frequently kept. Fishing was also important, but it did not become a major industry in Iceland until the late 19th century and was rarely carried on as an occupation totally separated from farming; fishing boats were generally owned by farmers who were better off and crewed by the poorer people.

Land Ownership

Icelandic society in the 18th century consisted primarily of farmers and tenant-farmers who were by any standards extremely poor. The census taken in 1703 (Magnússon and Vídalín 1924) records that a considerable proportion of the population consisted of landless labourers.

The system of land ownership made it difficult to acquire land and also acted as a disincentive to improvement. A large number of farms in Iceland were owned by the crown or church and these were let or sub-let to the peasant-farmers. Rents had been fixed at one twentieth of the (fixed) taxable value of the farm and no changes had been made in rents over two or more centuries. However, since the productivity of the farms had declined (due in part to natural disasters such as volcanic eruptions and bad weather conditions) a heavy burden was placed upon the peasants and they were frequently unable to pay. Those who were sub-lessees led a particularly difficult life as, in addition to paying high rents, they had to give various forms of service to the landlords such as rowing fishing boats during the fishing season or feeding a certain number of the landlord's cattle over winter, something that was very difficult during the frequent times of hay shortage (Jónsson, 1892-94).

Housing

The poverty of the peasant-farmers was reflected in the way they lived. Lacking timber and other building materials, their dwelling had to be constructed from the only materials available; turf and stone. Jónsson (1892-94, 602-603) describes the interior of such a home:

> *"The 'Baðstofa' (living room) usually did not have any boards covering the walls inside. One could see between the rafters right up to the grass-covered roof, ... and from which mildew and cobwebs hung down. The floor was also uncovered and made of earth stamped down but during heavy rains, against which the roof did not afford much protection, the water dropped down and the floor was soon turned to mud through which the people waded. The walls along which the bedsteads were nailed fast were covered with a grey coat of mildew... The poor people did not have many bed clothes, old hay, seaweed or twigs served as a mattress and a few blankets constituted the covering ..."*

Diet

The main food sources of Iceland consisted of fish, meat and milk products. However, peasants lacked butter and meat. Butter was frequently used to pay taxes to their landlords leaving little for themselves, and much of the meat produced was exported (Jónsson, 1892-94, 607). Frequently they had little fish for their own use either as they were forced to spend much of their time working in the King's or other landlords' fishing boats (Jónsson, 1892-94, 587). The peasants therefore had to supplement their diet as best they could with what they found growing wild. Those who lived inland gathered wild plants, especially Icelandic moss (*Lichen Islandicus*) and angelica roots which were generally cooked in milk, whilst those who lived by the sea used various types of seaweed. Some of the peasants were able to afford a little imported flour, a quarter to half a barrel per year (Aðils, 1926-27, 472) with which they made a kind of porridge or soup. (It is very likely that in the mid 18th century in Iceland there were many people who had never tasted bread.)

CLIMATE IN THE EIGHTEENTH CENTURY

During the 18th century Iceland experienced many exceptionally severe winters and cold and rainy summers. The Arctic sea ice was also a frequent visitor to Icelandic coasts. The severest years of the period occurred in the 1690s and early 1700s, in the 1750s, and again at the end of the century. The year 1695 stands out as being one of the coldest of the Little Ice Age in Iceland as well as in the rest of Europe. It was also one of the most severe sea ice years in Icelandic history, when the ice reached round the west and south coasts (Vilmundarson, 1972). In the 1750s another series of very harsh winters and cold summers occurred, reaching a climax in 1756. In this year the sheriff of Húnavatnssýsla in the northwest wrote:

> *"The sea was covered by ice the whole summer and this has prevented all fishing and all delivery of foodstuffs from Copenhagen and also affected the grass-crop. And it has not been possible to harvest the little which has grown because of the continual fog, sleet and rainy weather caused by the ice, together with a strange cold in the air ..."*
>
> (B. S. Húnavatnssýsla, 1756).

Some contemporaries believed that circumstances were more difficult then than they had been at the beginning of the century:

> *"The oldest men who remember the severe years ... at the beginning of this century, say they were not so severe in this area as there was not such a great mortality of livestock and people as now."*
>
> (B. S. Dalasýsla, 1757)

At the end of the 18th and beginning of the 19th centuries a similarly difficult time was experienced (Andrésson, 1964). Apart from these particular periods of severe weather there are numerous other references to very cold winters or unfavourable weather occurring throughout the century.

The severe years of the 1690s and early 1700s and the 1750s were marked by strong glacial advances and the majority of Icelandic glaciers reached their post-glacial maxima during the latter period. After a retreat from about 1760-1790 another minor advance occurred over the period 1790-1810. Documentary evidence provides us with some details concerning the movement of the glaciers and the abandonment of farms and farmlands due to the encroachment both in the northwest (Drangajökull) and the southeast (Vatnajökull). For example, in the land register from 1709 for Austur-Skaptafellssýsla, in the southeastern part of Iceland which is dominated by Vatnajökull, interesting details are given for two farms in particular (Fjall and Breiðámörk) and it is stated that these were derelict, having recently been encircled and destroyed by the glacier (Þorkelsson, 1918, 33-34).

CLIMATE AND FARMING

Eighteenth century farming practices in Iceland had not altered greatly since medieval times. As mentioned above, farming was based upon animal husbandry and life centered on the seasonal round and the natural cycle of the animals. During the summer months the sheep and some of the cattle and horses were let loose to fend for themselves on uncultivated common land to find what they could to eat in the way of grasses, sedges and small shrubs. In this way they could get maximum benefit from the frequently sparse vegetation. In winter most of the animals were kept inside to protect them from the cold, but also because a snow or ice cover would prevent grazing. For the animals to survive the winter, therefore, they had to be given fodder for several months between October and June, the exact period depending on the severity of the winter and spring. They were fed on hay. According to modern calculations, the winter feed requirements of one sheep is, on average, 3 cubic metres, one horse, 10-15 cubic metres, and one cow, 35-40 cubic metres. Eighteenth century animals probably had to make do on less. The farmers' most important summer task, therefore, was grass cutting to provide enough hay for winter feed. To this day, the main farm crop is still hay.

Hay was harvested from the "homefields" close to the farm and from the outlying pastures. Reaping was done by hand, usually in July and sometimes again at the end of August, depending on the stage of the growth and the weather. After it was cut the hay would be stacked or hung up to dry. The homefield was regularly manured and this was the only direct attempt made to improve the yield. This manuring was probably rather inefficient; a Faroese visitor to Iceland in 1780 described how manure dried out in the spring winds:

> *"In the spring the manure was spread out on the homefield where, in the strong winds it soon dried. A large part of it was thus reduced to dust and blew away, the rest formed into lumps and would be raked away when the grass began to grow."*
>
> (Mohr, 1786, 358-359 - translated from Danish)

A few agricultural improvements were begun in the late 18th century; such as the fencing in of certain pastures, and the attempt to flatten out hummocks (caused by frost action) which hindered the easy cutting of the hay. Hummock flattening was most laborious and even if a farmer spent much time trying to flatten them, they could reoccur after severe weather. More significant changes in agriculture did not occur, however, until much later and, prior to the late 19th century, grass-fields were not sown but just left to grow naturally.

Although grass-growth is influenced by a variety of non-climatic factors (soil type, length of the day, height above sea-level, etc.) climatic variables such as summer temperature and rainfall and the extent of frost in winter and spring are extremely important. The presence of sea-ice off the coast also has an adverse affect on grass-growth (Friðriksson, 1969).

The sensitivity of agriculture in 18th century Iceland to climate variations can be seen by reference to modern studies. One such study found that, allowing for the influence of fertilizers, the hay-yield in the period 1931-60 was 16 per cent higher than it would have been if weather conditions had been like those of 1873-1922. Also the time of winter feeding for cattle was 13 days sorter (September 27 to June 25 in 1873-1922; September 30 to June 15 in 1931-60). In the later period a farmer could keep 20 per cent more cows, 30 per cent more sheep and 30 per cent more horses than in the earlier years assuming that the size of his homefield and his methods of farming remained the same (Bergþórsson, 1966). Conditions during most of the 18th century were significantly worse than for the period 1873-1922.

It is obvious that in northern latitudes any kind of agriculture is precarious. The difficulties are increased in Iceland by the great variability of the climate. The formula for disaster in earlier centuries was simple: if there was not sufficient hay to feed the livestock over the winter, the animals would, at best, become so emaciated and weak that they would be of little use - at worst, they would die of hunger. Deprived of their livelihood, the people also faced starvation and death.

There were many different sets of climatic circumstances which could have severe effects on the people of Iceland. For example:

1) A severe winter with snow and ice which lasted into the spring and early summer delayed the growth of the grass;
2) Rain during the harvest or when the grass was reaped could cause it to spoil. However, insufficient rain meant that the grass did not grow well;
3) The presence of sea-ice off the coast also affected the grass crop, both because of lowered temperatures and because the fog and rainy weather often associated with it could hinder the harvest. Sea-ice is most frequently to be found off the coast of Iceland from late winter to early spring. If it stays longer, into the growing season as in severe years, the length of the growing season is reduced.

A combination of these circumstances, such as a rainy summer followed by a severe winter, spelt disaster. If it looked as though it would be impossible to feed all the livestock through the winter many of them were slaughtered in the autumn. People were often reluctant to do this, however, and invariably many of the cattle died of starvation. For example, during the 36 years between 1730 and 1766, an analysis of primary sources shows that livestock died from hunger and cold in at least 24 of these years. During the severe years of the 1750s the problem was particularly acute. A typical letter of the time was written in 1757 by Magnús Ketilsson, sheriff of Dalasýsla:

> *"... the conditions in this district ... are so difficult that if I described them without furnishing clear proof they would seem incredible. I have therefore found it necessary to note it in official Assembly testimonies and I enclose a copy of each one for your Noble Excellency's perusal. With regard to these testimonies I should like to ... draw your ... attention to the fact that just in this year 321 cows and bulls, 1292 sheep, 3209 young lambs and 151 horses have died in this district. Forty-four people have died of hunger and wretchedness, and 15 dwellings have been deserted. Furthermore ... during the year 1755 far more cattle died, in some places more than three times as many as during the previous year. With this loss of livestock and the resulting difficulties it is a foregone conclusion that during the coming winter there will be such dearth and misery in the district that it will far exceed the dearth that has already been experienced. And it will be a pure miracle if a third of the population does not die of hunger as the small number of livestock which the peasants have left over is so wretched and useless that they have no milk. They do not have any fish either as the fishing off the coasts the previous spring and winter was poor and they have no cattle to slaughter, with which to maintain life."*
>
> (B. S. Dalasýsla, 1757)

CLIMATE AND FISHING

Since medieval times Icelandic waters have been renowned as rich fishing grounds and fish have formed a staple part of the Icelanders' diet from the time of their first settlement there at the end of the 9th century. From the 14th century onwards fish was the main export item. The most important catch was cod and this was the favoured type of fish both for export and home consumption.

As the people only had small, open rowing boats and they did not venture very far out to sea, the catches were never very large. There is, however, considerable evidence that, during the 18th century, fishing catches declined sharply. Most contemporary writers seem to have noticed this fact and remarked upon it. When 18th century catches are compared with those of the 17th century in the south of Iceland one man would be expected to catch, on average, about 500 cod during the winter fishing season, but, after the mid 18th century, 300 cod were regarded as the maximum catch (Aðils, 1926-27, 515). Also, in the 18th century the number of fish exported fell and the poor catches are reflected in the reports of Icelandic officials. For example, a report from Barðastrand district in the west, for 1747, reads:

> *"In the west, the fishing was poor; there were average to small catches of lump fish but everywhere the cod catch was the worst that people could remember for many years, both during the autumn fishing last year and this spring. In most places most boats did not catch more than about 10 fish and in many places, less."*
>
> (B. S. Barðastrandarsýsla, 1747)

The failure of the fisheries, which lasted from the late 17th to the late 18th centuries, seems to have peaked around the periods 1685-1704 and 1744-1759 (Kristjánsson, 1971a; Gunnarsson, 1979).

There are several possible explanations for the poor catches of fish in 18th century Iceland. When fish disappeared from some fishing grounds people who had lived in these areas and were dependent on fishing had to leave their homes and try to find a living elsewhere. This, plus the great smallpox epidemic of 1707-08, which wiped out one third of Iceland's population, meant that farmers living in the coastal regions had great difficulty in getting sufficient people to man their boats (Aðils, 1926-27, 516-517). This view is supported by contemporary evidence as the following example shows:

> *"Instead of the usual 14 or 15 boats from Ólafsvík only 2 could be manned this year because of the lack of labour. Many people who lived by the coast have left and the people who live further inland do not want to risk setting out to fish because of the recent failure of the fishing."*
>
> (B. S. Húnavatnssýsla, 1732)

The lack of labour meant that farmers had to build smaller ships. In the 17th century boats with 10 or 12 oars were most usual, but even before the middle of the 18th century boats for 6 men or even smaller came into use. This, apart from anything else, would mean smaller total catches. However, the main reason for the decline in fishing catches appears to be related to climate, more specifically to sea temperatures and ocean currents. This is because of the influence of these factors on the development and spread of fish species. As stated above, cod was the main Icelandic catch and, although the exact relationships are by no means clear-cut, it seems that cod needs temperatures of 4 to 7°C for reproduction and early survival (Woodhead and Woodhead, 1959). The growth and decline of the Greenland cod industry during the 20th century is partly the result of changes in sea temperatures. Early this century, temperatures rose and the Irminger current increased in strength. This meant that the cod species was able to expand its range further north in the Atlantic Ocean (Beverton and Lee, 1965) and that cod transported from Icelandic waters by ocean currents began to breed in Greenland's waters, allowing the development of a Greenland cod industry. When sea temperatures cooled again later in the century the fishing industry declined; the water had become too cold for the cod to breed.

It would seem likely, therefore, that during a much colder period, as in the 18th century, the range of the cod would have decreased and withdrawn southwards. The presence of the Arctic sea ice would have had a doubly detrimental effect on fishing in the 18th century; both because of its effect of lowering temperatures and because its presence off the coast prevented people from going out to fish. This suggests a good correlation between sea ice extent, sea temperatures and ocean currents and cod development, and hence between severe weather and failure of the fisheries. A comparison of the 18th century climatic record from Iceland with fisheries records supports this (Gunnarsson, 1979).

As mentioned above, the worst years for fishing during this period were 1685-1704 and also around 1744-57. Climatic conditions during these years were also exceptionally severe. Complaints of poor catches did, however, occur throughout the 18th century. To take just one district, that of Skagafjord in the north, the letters from the sheriff have these comments to make on the state of the fishing:

> *"1729: occurred late because of the sea ice; 1730: very poor; 1731: failed here in the north; 1732: very poor; 1773: the fish seem not to have come as close to the land as usual. This is believed to be because of the drift ice; 1734: failed in some places; 1735: poor; 1738: scarcely any fishing; 1749: very poor; 1751: poor; 1754: lack of fishing; 1758: very poor."*
>
> (B. S. Skagafjarðarsýsla, 1729-1758)

Climate was of importance for the fishing in other respects as well. If the weather was particularly stormy then many lives were lost at sea, and more fishermen were drowned during cold and stormy periods (as in 1698 to 1704: Kristjánsson, 1971b). The weather conditions were also important *after* the fish had been caught. Icelanders had always dried fish instead of salting it and it was primarily in this form that it was both exported and consumed. Fish that was caught during the winter was gutted and the backbone taken out. After that it was left in heaps where it lay untouched until after about mid-April. The fish caught after that time was cleaned and dried immediately. The fish was generally dried in the open (the lack of timber prevented the building of drying sheds). If the weather remained constantly cold or dry the fish would remain in good condition. In variable or wet weather, however, the fish tended to deteriorate and become unsuitable for export (Aðils, 1926-27, 517).

FURTHER EFFECTS OF SEVERE SEASONS

The effect of the greater frequency of severe seasons in the 18th century on the two traditional foundations of the Icelandic economy, farming and fishing, can be seen clearly: in both cases the primary effect was a decrease in food supply. This had several further consequences; for example, farms were deserted when people left their homes in order to go and beg for food elsewhere, crime increased as people became more desperate and saw stealing as the only means of survival, cultural life was affected adversely, and, in the most difficult years, people died of starvation. All these problems were inter-related and the difficulties affected the whole society. When farms were deserted many people who worked for others lost their jobs and security. Also, as the poorer people died, there were complaints from the more well-to-do of the difficulty of getting servants. Even the Danish merchants suffered when people were unable to pay their debts or their trading vessels were blocked by sea-ice.

Deserted Farms

It is likely that the desertion of farms was a widespread phenomenon throughout the 18th century. However, different districts were not necessarily affected in the same way. Economic and climatic conditions varied to some extent from place to place, but, on the whole, the poorer districts were those situated in the north and east and these were often the first to be affected during a severe season. Thus, it is not surprising that most of the documentary evidence suggests that the pattern of abandonment involved primarily people from the north and northeast travelling to the western fjords and to the southwest. In the very severe year of 1756 the sheriff of Húnavatnssýsla wrote:

> *"In this district during the winter and spring ... many left the farms on which they lived and sought help and charity in other parts of the country."*
>
> (B. S. Húnavatnssýsla, 1756)

Sometimes the resources of a district that might not otherwise have been too badly affected were put under severe strain by in-migration. In 1756 David Schieving, sheriff of the district of Barðastrand in the west, said:

> *"... what is proving the ruin of our inhabitants more than anything else is the great number of people coming from the northern part of the country as these take the little we have, and it already seems as though the situation will soon not be much better than in the north."*
>
> (B. S. Barðastrandarsýsla, 1756)

Many of those who did leave their homes must have found that there was little help to be found in other districts. A letter from 1732 from Snæfellsnes, supposedly one of the better-off districts, illustrates this:

> *"... During the severity of the previous winter many people have come here to this fishing place from 20 and 30 miles away but they have*

had little fish to eat ... nearly all the tenant farmers and crofters who live by the sea have left their dwellings to go begging for themselves and their children."

Some other sources are blunter in their descriptions. For example, one annal states for the year 1755:

"There was a lot of traffic of people who had left their homes, especially from the north, which increased after the previous severe winter. That spring all over the country poor people died of hunger, in poverty, and some were found like cattle out on the heaths."

(Sauðlauksdalsannáll)

Crime

There is little doubt that the crime-rate rose sharply during severe seasons. Many of the "crimes" mentioned during difficult times, such as the stealing of food and other goods, reflect the situation of the people, hungry and desperate as they were. During the severe years in the 1750s the amount of crime and lawlessness seems to rise gradually from about 1751 and reach a peak in 1756, the most difficult year of that period. The punishment for stealing even a small amount of food was harsh and could include flogging, branding, and being condemned to a life of hard labour at the fortress Bremerholm in Copenhagen. The lack of any proper facilities in Iceland for people regarded as criminals was acutely felt during such times and those who were to be kept prisoner had to be sent to Denmark. This problem was exacerbated by a law passed in Denmark in 1751 which stipulated that certain criminals should be given hard labour for life. During the 1750s there were many pleas that a prison might be built in Iceland (although this was, to some extent, seen more as a poor house where the large number of beggers could be accommodated) and finally in 1759 the King recommended that this be done.

Mortality

The population of Iceland fluctuated considerably during the 18th century and the high mortality rates and low birth rates during certain years or periods has a clear link with the effects of the severe weather which caused failure of the grass and the deaths of livestock. Although a large number of people died of starvation, many also died from diseases which were related to hunger and malnutrition, such as dysentery and gastero-enteritis. In addition to this, infectious diseases such as smallpox and influenza spread more rapidly through a population that was already weak.

The first census in Iceland was taken in 1703 when the population numbered 50,358. In 1707 this was reduced by about one-third by a severe smallpox epidemic. In 1734 the population was 43,377. It rose to around 48,000 by 1755, but by 1759, after the severe seasons of 1756 and 1757, had dropped to 42,822. By 1778 the population had recovered to 49,863. Severe seasons again occurred in the late 1770s and early 1780s, and in 1783 there was a devastating series of volcanic eruptions. By 1786 population had dropped to 38,363.

From about 1735 there are quite good annual data on births and deaths. These do not always make it clear what the specific causes of death were, but the death-rate certainly peaks during the climatically more severe years of the century; for example, during 1739, 1742, 1752, 1756-58, 1763, 1767, 1782 and 1784-86. The highest death-rate figures are for 1756-58, years of extremely severe weather, and for 1784-86. These latter years *were* severe climatically but the main difficulties here were associated with the eruption of the volcano Laki in 1783. During the severe years 1750-1758, it has been calculated, nearly 17,000 people died. (These population data are to be found in the various publications of the Statistics Bureau of Iceland, and in Imhof, 1976.)

Clearly the populace was greatly weakened during severe years. In Barðastrandarsýsla the sheriff's report for 1748 states that

> *"... the people suffered from lack of foodstuffs having no milk and little fish and they were forced to use their remaining livestock for food if they were not to die or be forced to beg."*
>
> (B. S. Barðastrandarsýsla, 1748)

In 1749 we read:

> *"... there has been sickness in the district in the autumn (1748) winter and spring and many people have been in bed with pains in the head, legs and limbs, and some have died ... the lack of food after last year has caused suffering to many, especially in the southern part of the district."*
>
> (B. S. Barðastrandarsýsla, 1749)

In 1751 there was a very cold spring, the grass growth was poor and so was the fishing. The sheriff does not mention the state of the people then, but in 1752, when there was a tolerable winter but another very severe spring, he writes:

> *"As a result of the rain and frosts in the early spring the ground became ice-covered and there was nothing for the livestock to eat so they had to be given hay, but this was in short supply after last year's meagre harvest. The livestock, especially sheep and horses, died in great numbers. People also died from lack of food."*
>
> (B. S. Barðastrandarsýsla, 1752)

Sometimes the sources give information that is a little more precise. For example, in 1756 in the district of Þingeyjar on the church farm of Helgastaðir, four people were said to have died, "two of hunger"; in 1757 seven died, "six of hunger"; and in 1785 nine died, "eight of hunger" (Imhof, 1976, 279).

There was little doubt in the minds of contemporaries as to the causes of the increase in deaths during the severe seasons. In 1757 the sheriff of Dalasýsla wrote:

> *"The main cause is given as the lack of grass and the poor harvest last year. People have become so impoverished that they would be unable to buy foodstuffs even if they were readily available. The people who have died of hunger, including those who have come from other settlements, number seven in all. Two farms have been deserted."*
>
> (B. S. Dalasýsla, 1757)

Here he is only referring to the immediate vicinity in which he lived. In that entire district there were at least 54 deaths that year and 30 farms were deserted: Table 1 gives further details from this particular source:

TABLE 1

Deaths of Livestock and People, and Desertion of Farms: Dalasýsla, 1757

	Cattle	Sheep	Lambs	Horses	People	Farms Deserted
STAÐARFELL	29	117	322	11	7	2
HVAMMUR	35	182	474	21	7	1
HJARÐARHOLT	55	236	500	31	7	2
JÖRFI	21	80	372	6	7	-
SAUÐAFELL	73	205	455	21	7	2
BLÖNDUHLID	42	104	273	2	8	5
BALLARÁ	24	160	242	30	1	2
STAÐARHOLT	42	263	565	24	10	16
TOTAL	301	1347	3203	196	54	30

CONCLUSION

Historical events and circumstances are rarely monocausal and those of the 18th century in Iceland are no exception. The Danish trade monopoly, the system of land tenure, disease, the frequency of severe seasons and other less well-defined factors all combined to bring about the difficult socio-economic conditions described in this paper.

The limited Icelandic economy, based on animal husbandry, with supplementary fishing, increased the country's vulnerability to the vagaries of climate. The farmers' main task was always to harvest enough grass during the summer to feed the livestock during the winter. In the 18th century, the harvest often failed, due either to excessive rains during the summer or to a cold spring when the grass did not grow until very late, or to a combination of these. If a harvest failure was followed by a severe winter with prolonged snow and ice cover, the livestock frequently starved to death. The presence of sea ice during severe years, particularly off the northern coast of Iceland, prevented sea-fishing and destroyed shore vegetation and generally exacerbated the situation. In exceptionally severe years, or when severe weather lasted for several years, as it did in the 1750s, there was also human mortality.

Climatic stress affected all of society, but its influences were most obvious for, the most detrimental to, the peasants. Poor harvests and severe winters caused many to desert the farms and migrate to other districts, where conditions were often little better. Some turned, in desparation, to crime and many, like the livestock they owned, died. Land-owners were perhaps less severely affected. The poor harvests did affect them directly, of course, but they, and others of the more well-to-do were, also affected by the lack of labour and servants.

How did society adapt to the harsher climatic conditions of the 18th century? The evidence suggests that it did not. The peasants had neither means nor opportunity to adapt or modify their agricultural methods, and were unable to divert extra energies to fishing. The latter possibility was denied them, anyhow, because a decline in fish stocks seems to have occurred during the climatically severe periods. Land-owners made no significant improvements in agricultural practices, and changes in fishing were dictated partly by the diminished fish stocks and partly by the reduced labour pool. It is true that the land-owners were less affected than the peasants by the severe conditions and so had less pressure to adapt. However, vested external interests were also important, and Danish merchants, on whose trade the Icelanders' depended, would have favoured maintaining the existing economic and social structure.

In many ways Iceland is ideal for the study of the impact of short and medium term climatic change on society - a country where settlement and agriculture are marginal, and where climate is highly variable on all time scales. Climatic impact is obvious during the 18th century, but its effects were strongly modified by other factors. Significantly, these same factors are important in explaining why the society adapted so little in the face of such climatic stress.

REFERENCES

Manuscript Sources

National Archives, Reykjavík

Bréf til Stiftamtmanns úr Barðastrandarsýslu 1709-1785
Bréf til Stiftamtmanns úr Dalasýslu 1704-1785
Bréf til Stiftamtmanns úr Húnavatnssýslu 1711-1735, 1736-1792
Bréf til Stiftamtmanns úr Skagafjarðarsyslu 1708-1740, 1741-1790
Bréf til Stiftamtmanns úr Snæfellsnessýslu 1711-1754

National Library, Reykjavík

Sauðlauksdalsannáll, Lbs. 494, 4to

Published sources

Aðils, J. J. (1926-27) *Den Danske Monopolhandel På Island 1602-1787*. Copenhagen.

Andrésson, S. F. (1964) Harðindi á Íslandi 1800-1803. *Skírnir*, 138, 5-44.

Bergþórsson, P. (1966) Hitafar og búsæld á Íslandi. *Veðrið*, 11, 15-20

Bergþórsson, P. (1969) An Estimate of Drift Ice and Temperature in Iceland in 1000 years. *Jökull*, 19, 94-101.

Beverton, R. J. H. and Lee, A. J. (1965) Hydrographic fluctuations in the North Atlantic Ocean and some biological consequences. In C. G. Johnson and

L. P. Smith (eds.) *The Biological Significance of Climatic Changes in Britain.* Academic Press, London, 79-107.

Friðriksson, S. (1969) The Effects of Sea Ice on Flora, Fauna and Agriculture. *Jökull*, 19, 146-157.

Gunnarsson, G. (1979) *Fishery Records in Iceland and Sea Temperature in Icelandic Waters during the Little Ice Age.* (unpublished).

Iceland, 874-1974 (1975) Handbook published by the Central Bank of Iceland. J. Nordal and V. Kristinsson (eds.), Reykjavík.

Imhof, A. E. (1976) *Aspekte der Bevölkerungsentwicklung in den nordischen Ländern 1720-1750.* 2 vols, Francke-Verlag, Bern.

Jónsson, J. (1892-94) Fæstebondens Kår på Island i det 18 Århundrede. *(Dansk) Historisk Tidsskrift*, 7 vol. 4, 563-645.

Kristjánsson, L. (1971a) Úr heimildahandraða seytjándu og átjándu aldar þá eru komnir prír í hlut. *Saga*, 9, 123-139.

Kristjánsson, L. (1971b) Sjóslysaárin Miklu. *Saga*, 9, 158-170.

Magnússon, Á. and Vídalín, P. (1924-8) *Manntal á Íslandi árið 1703.* Reykjavík.

Mohr, N. (1786) *Forsøg til en Islandsk Naturhistorie med adskillige oekonomiske Samt andre Anmærkninger.* Copenhagen.

Parry, M. L. (1978) *Climatic change, agriculture and settlement.* Dawson Press, Folkestone.

Vilmundarson, P. (1972) *Evaluation of Historical Sources on Sea Ice near Iceland.* Reprinted from Sea Ice Conference Proceedings, Reykjavík, 159-169.

Woodhead, P. M. J. and Woodhead, A. D. (1959) The effects of low temperatures on the physiology and distribution of cod, *Gadus morhua L.*, in the Barents Sea. *Proceedings of the Zoological Society, London*, 133, 181-199.

Þorkelsson, J. (1918) Jarðabók Ísleifs Sýslumanns Einarssonar Yfir Austur-Skaptafellssýslu Árna Magnússonar 1708-1709. *Blanda*, 1, 1-38.

TAX RECORDS AS AN INDEX OF LITTLE ICE AGE ENVIRONMENTAL AND ECONOMIC DETERIORATION, FROM SUNNFJORD FOGDERI, WESTERN NORWAY 1667-1815

Jean M. Grove
University of Cambridge

Arthur Battagel
Pulborough, Sussex

Data from general tax commissions held in Sunnfjord Fogderi reveal a substantial decline in rural prosperity between 1667 and 1723. Late seventeenth century and eighteenth century incidence of serious physical damage to farmlands is documented in tax relief proceedings. Environmental deterioration characterised the early years of the Little Ice Age in western Norway.

The glacial advances of the Little Ice Age in Norway appear to have occurred about a century after those in the Alps. The earliest reference found to direct damage by ice from Jostedalsbreen, the largest and best documented ice mass in Norway comes from Kjendalsbreen. We learn from the report of an investigation following a claim for tax relief in 1693 of "A valley called Tiördahl, (Kjenndalen).....[which] has now been wholly lost and completely ruined by the destructive glacier and by the great river running from it, so that where they formerly had five enclosures and six barns, they and their contents have been completely carried away". General advances followed, the maximum positions of many tongues being reached in the 1740s. Retreat has predominated since about 1870, although this has not been without minor interruptions and stationary periods.

It has long been recognised that farms in the immediate vicinity of glaciers suffered serious damage during the period of ice expansion (Rekstad, 1902; Hoel and Werenskiold, 1962) but little attention has been devoted to investigation of the general extent of the environmental deterioration which must be assumed to have been associated with it, nor to measurement of consequent economic effects. Fortunately, Norwegian tax and land rent records from the seventeenth and eighteenth centuries are voluminous and have characteristics which make them a peculiarly fruitful source of data for investigations of this type.

NORWEGIAN TAX AND LAND RENT RECORDS

Landskyld or land rent was a key element in the Norwegian rural economy in early modern times (Holmsen, 1956, 1958). Before the Black Death it had provided a fixed value for a piece of land when it was bought, sold or inherited, and was also employed by the tax authorities as a measure

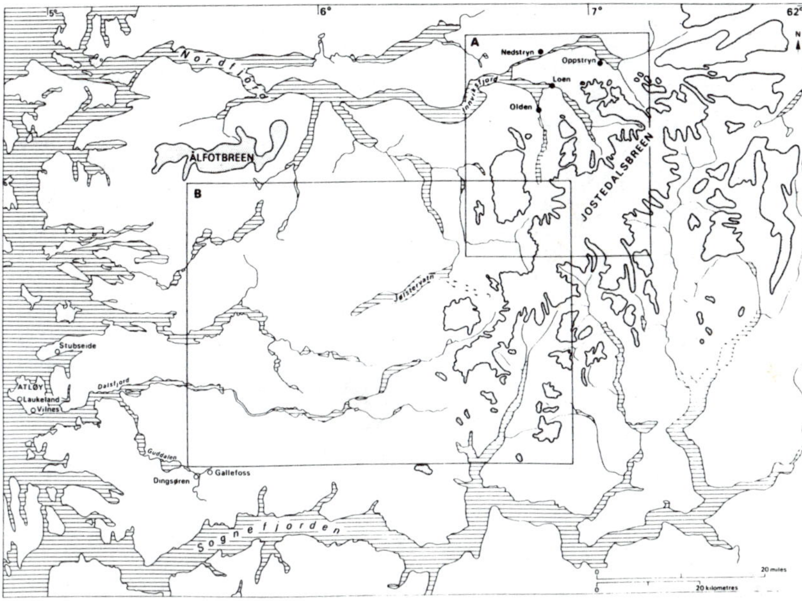

Figure 1 - Sunnfjord Fogderi; sketch map to show position of the fogderi in relation to Nordfjord, Sognfjord and Jostedalsbreen. Inset A - see Figure 3; Inset B - see Figure 9.

of taxability of all farms, even those of freeholders. The landskyld, which was assessed separately for each farm and for each of its constituent parts or *bruk*, was paid mainly or entirely in kind for many centuries. The nature and amount of the landskyld payment therefore reflected the nature of the local economy.

From the middle of the sixteenth century onwards the landskyld could not be changed without an official assessment. According to an Order in Council confirmed by Christian III published in 1539 "Concerning the payment of Landskyld or *leding*, these will be paid in corn, hides, slaughtered beasts or skins, in proportion to the produce of the soil, and after the pronouncement of judgement of six men" (Paus, 1751, p.292-3). Originally the value was officially fixed at one sixth of a farm's gross yield. Although there was a transition towards payment in money during the second half of the seventeenth century and by a decree of 4th March, 1684, tenants were given the choice of paying in cash or in kind, if cash was chosen the amount had to be the official value of the landskyld in kind. General tax or Matrikkel Commissions were held in 1665-70 and in 1722-24 to set the rate of assessments for wide areas or for the whole country.

SUNNFJORD FOGDERI

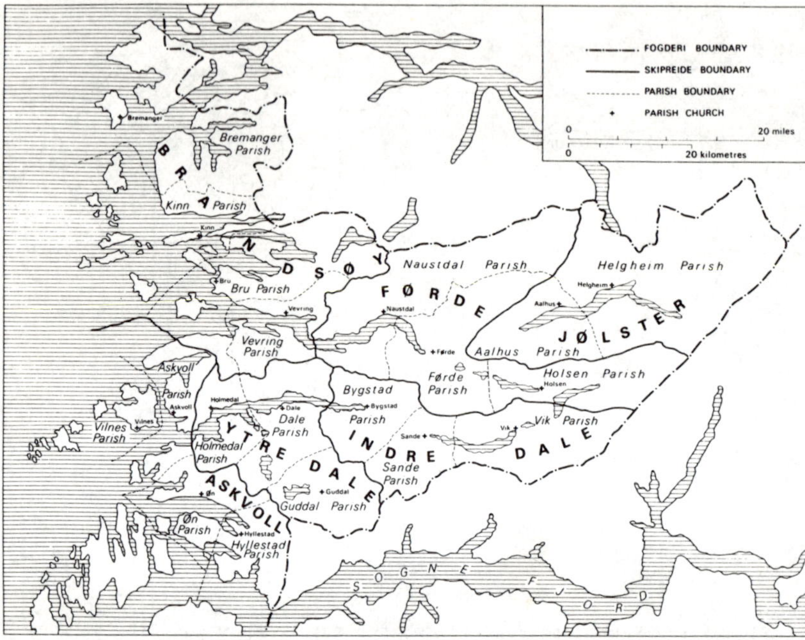

Figure 2 - Sunnfjord Fogderi - Administrative divisions

Sunnfjord Fogderi provides a convenient cross-section of environments between Jostedalsbreen and the coast between Nordfjord and Sognefjord, round about 61.5°N (see Fig. 1). A *fogderi* was an administrative division which may be taken as equivalent to an English county. This large mountainous area had a population of a mere eleven or twelve thousand in the first half of the seventeenth century. In Sunnfjord, as everywhere else in Norway, the population has fallen disastrously after 1350. This demographic collapse has been attributed in varying degree to the effects of the Black Death, climatic change causing declining yields in the fourteenth century, and variations in the amount of fish caught off the west coast. The first complete census did not take place until 1801, but according to recent estimates the total population was of the order of 180,000 in 1520 and 450,000 in 1665 (Hovstad, 1971). In western Norway resettlement spread from the west, up the fjords and then into the higher lying valleys. Accordingly in Jolster Skipreide, in eastern Sunnfjord (Fig. 2), only two farms were registered in the 1520 tax records but 37 containing 52 holdings or bruk in 1563. By 1590 there were 53, containing 95 bruk, while in 1631 there were 60 farms with 238 bruk. The real expansion was now over, for in 1647 there were 59 farms with 242 bruk and in 1667, 61 farms with 277 bruk.

In 1667 there were in all 774 farms scattered over Sunnfjord Fogderi. Very few of them were at altitudes over 200m and none over 400m. There were no villages as such; the farms were divided into a varying number of bruk, each with its own buildings. In this part of Norway the main buildings of all the bruk were huddled together in a *klyngetun* or cluster, to give a hamlet-like settlement. Returns off arable land were meagre. The people of Dversdal on Jølstervatn, for example, sowed 22 tønner of oats in 1667 and harvested 81.5 tønner, but this fourfold return was above average for the area. Some farms were getting less than a threefold return. The main dependence was upon cattle. This is nicely reflected in the Dversdal assessment of 1667 of 3.5 lauper of butter, 27 merker of tallow and 4 goathides. (A laup of butter, which was divided into 3 pund each containing 24 merker, was officially equivalent to 18 kg weight of butter, or two hides, four sheep or two riksdaler in coin in the Sunnfjord region. One and a half lauper was equivalent to a cow; so a laup was a very substantial unit of account). The pasture lands were the main productive element and hay the crop that really mattered. The vital question was how many cattle could be overwintered. The fate of the Sunnfjord farms and their cattle is revealed with some clarity in the tax and landskyld records.

A matrikkel commission operated in western Norway in 1667, and another followed in 1723. Both commissions operated in the whole country except for Finnmark. This involved unusually searching enquiries, for on this occasion the commissions were instructed to provide a detailed account of the whole rural economy. Not only were former and projected assessments recorded for each farm, together with details of the number of bruk, but also information about its situation and suitability for crops, its woods and pastures, fisheries and mills, horses and livestock and the amount of grain sown and harvested during the previous few years. Unfortunately the item on the hay crop was left blank for the Sunnfjord farms but in all thirteen other types of data were recorded, so providing a Domesday-like survey of the whole fogderi. As it happens this vast operation was abortive, for its recommendations were never implemented, but the documents, which survive in full, provide an excellent opportunity to compare conditions in Sunnfjord before the glacier fronts advanced into the valleys and in 1723 when very rapid expansion was taking place. It will be convenient to extend the sample area to take in the six parishes north of Jostedalsbreen which were part of the adjacent fogderi of Nordfjord.

Sunnfjord in 1667 and 1723: The cattle population

The number of cattle kept in each farm was recorded both in 1667 and 1723. It is most convenient to survey the resultant data on a parish basis. The administrative boundaries of Sunnfjord and the nearby part of Nordfjord Fogderi are shown on Figures 2 and 3 (each fogderi was divided into a number of skipreider, approximating to the English hundred, containing two or more parishes).

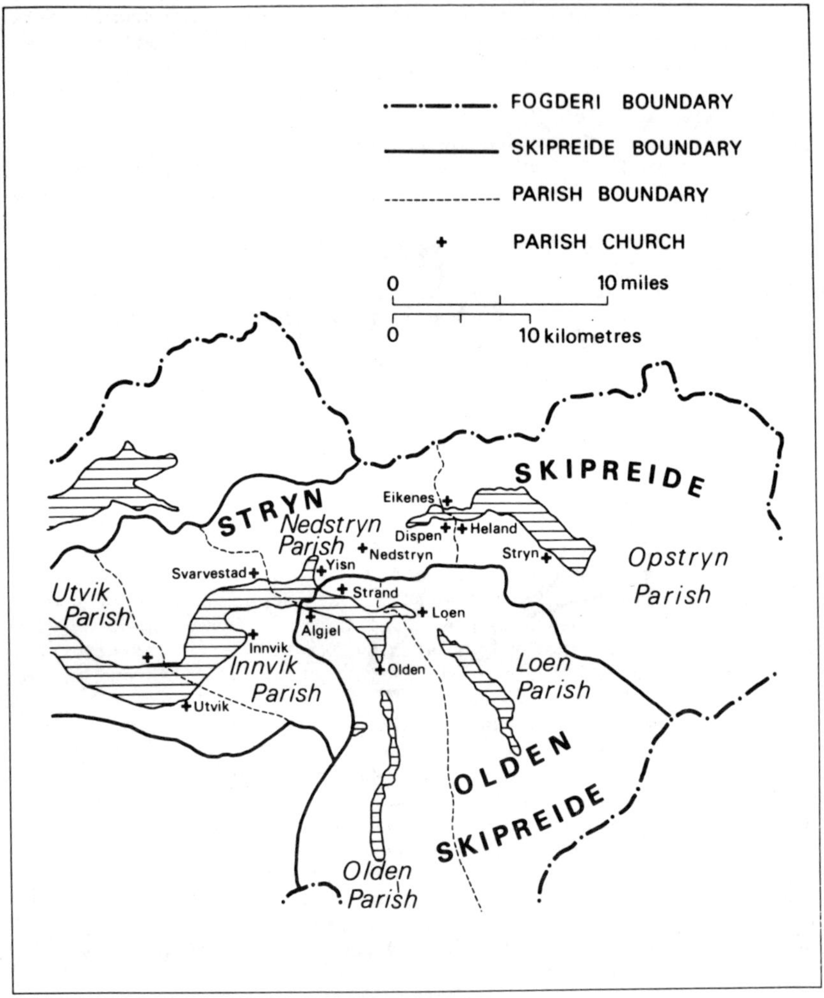

Figure 3 - Nordfjord Fogderi - Administrative divisions

Every parish in the whole region suffered a drop in the head of cattle kept between 1667 and 1723 and in most cases this drop was substantial. The percentage drop for each parish in Sunnfjord is shown in Figure 4. The diminution was most marked in a broad belt reaching from Jølster in the east through to Askvoll and Øn parishes in Askvoll Skipreide on the

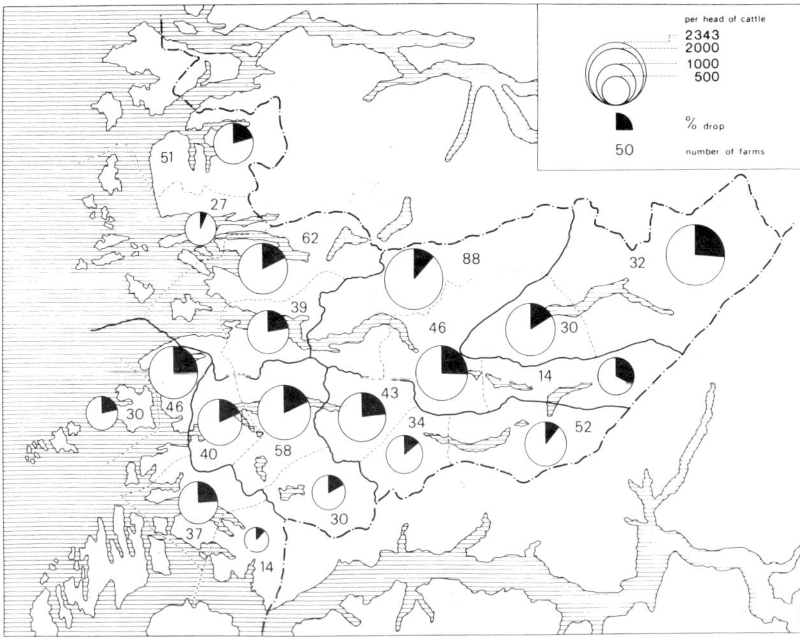

Figure 4 - The percentage drop in head of cattle kept between 1667 and 1723 in the parishes of Sunnfjord.

coast. Of the 19 parishes in the fogderi, only four near the northern boundary, and two, Naustdal and Kinn near the northern boundary, showed a loss of less than 15 per cent. Eight showed a drop of more than 20 per cent. Although the two most badly affected parishes, Holsen with a loss of 33 per cent and Helgheim with 26 per cent were in the east near Jostedalsbreen, the pattern of economic deterioration suggested shows little sign of amelioration towards the coast. Moreover, it is noteworthy that parishes with the best pasturelands were amongst the worst hit. Kraft (1830, p. 847-9) described Jølster as having "the best pastures in the district", followed by Holsen parish and Haukedalen, both in Førde Skipreide.

The position in the Nordfjord parishes was even worse; five out of the six had lost more than 20 per cent of their cattle by 1723 (Fig. 5). The most spectacular change in the number of cattle kept was the 39 per cent drop in Olden. Significantly, the farms in this parish had no saeters for they were practically restricted to the floor of steep-sided Oldendal. One of the few exceptions was Tungøyane, a farm which, like its neighbour Aabrekke, used summer pastures in Brenndalen, a small

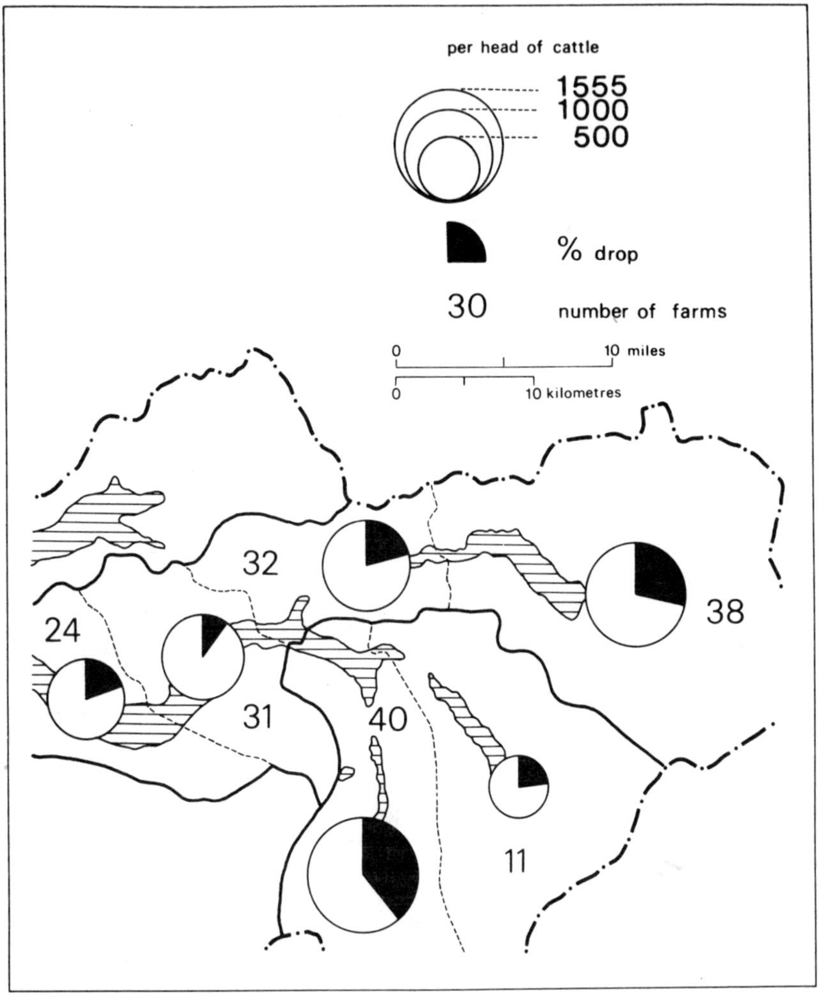

Figure 5 - The percentage drop in head of cattle kept between 1667 and 1723 in the parishes of Stryn and Olden Skipreider, Nordfjord Fogderi.

tributary valley. But ice from Jostedalsbreen was already accumulating in upper Brenndalen by 1685. By 1732 it filled the whole of the valley and the front of the glacier hung over Tungøyane. Finally the farmhouse itself was obliterated in December 1743 by falling ice and rock debris (Rekstad 1901). The other farms in Oldendal were not exempt from catastrophe, as will be seen below.

There is no doubt that the returns for cattle kept in 1667 and 1723 strongly suggest economic decline. It is necessary however to consider whether the data are sufficiently reliable for this decline to be taken as real and if so whether there is evidence to reveal the nature of the causative factors, either environmental or socio-economic.

Figures produced for the use of tax officials must always be suspect but it is worth noticing that although cattle were the mainstay in Sunnfjord, a diminution in the number of cattle which could be kept was not accepted as reason for reducing landskyld or tax assessments, except in special circumstances to be discussed below. Farmers would not therefore have been tempted to reduce their returns in the same direct way as would have been the case in, for example, 1657 when a special cattle census was held in connection with the imposition of a temporary cattle tax per se (this was withdrawn before 1667, Sandnes 1971).

The 1667 commissions were instructed to carry out their survey as soon as practicable after the harvest. Oats could not be harvested in Sunnfjord until mid-September near the coast and so we may presume that the work would not have begun before October. The 1723 commission examined Stryn and Olden Skipreide between 11th and 18th October and worked between November 2nd and December 15th in Sunnfjord. Therefore both commissions took place in the autumn and it is reasonable to take it that the winter herds not the summer herds were enumerated. The figures for individual farms seem to have referred to an average, as it was often noted that "in good years" more could be kept. It appears from the internal evidence that the figures should be taken as averages for a few years preceding 1667 and 1723.

The 1667 commissioners were instructed to examine for themselves any farm for which they had reason to doubt the veracity of the returns made. The commissions were accompanied by local assessors drawn from each skipreide in turn who would not have been likely, as a group, to allow individual farmers to get away with returns far different from the norm. In 1723 the commissioners certainly saw the farms for themselves. Their exact route is known (Fig. 6) and the procedure was far too public for it to have been possible for these remarkably hardworking officials to have taken short cuts or departed from the procedure laid down.

Sandnes (1971, p.308-9) surveyed the rather scanty literature available on the validity of cattle returns in Norway. The only detailed modern investigation he quotes is that made by Kåre Lunden, who compared cattle returns made in a 1668 matrikkel for Nordre Gudbrandsdalen farm by farm with contemporary deeds of conveyance and came to a provisional conclusion that the returns were possibly 15 per cent below the true level.

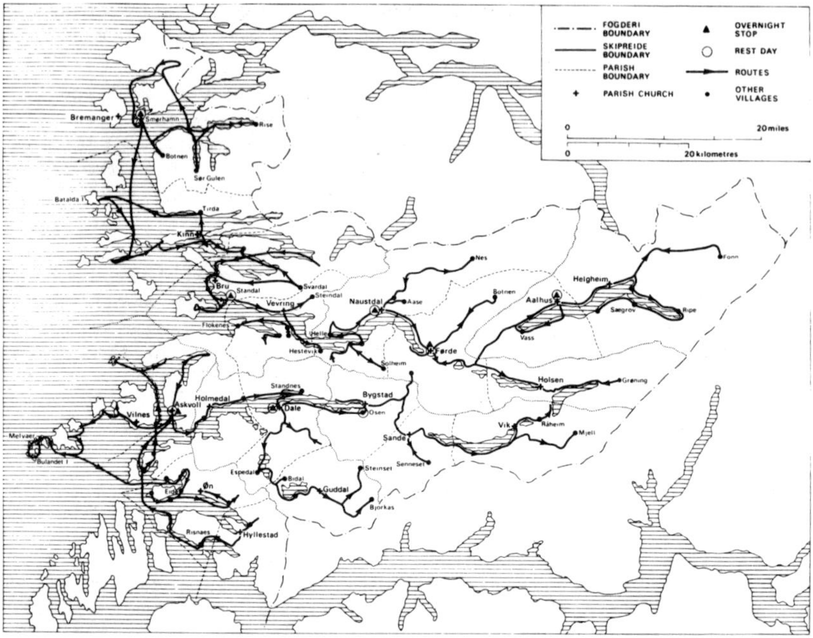

Figure 6 - The route of the 1723 Commissioners through Sunnfjord Fogderi

The cattle returns from 1667 and 1723 cannot be taken as precise but it appears reasonable to accept them as estimates to be taken sufficiently seriously to make it worthwhile looking for supporting evidence of deteriorating conditions. This is not far to seek.

Sunnfjord in 1667 and 1723: Physical damage to farms

The 1723 commissioners found that a sizeable minority of the farms they visited were in obvious danger of physical damage. Such information would have been of considerable fiscal interest, for the legal system had built into it a provision to meet the incidence of major physical hazards. A petition might be made for landskyld and tax relief if a farm was physically damaged by landslides, rockfalls, avalanches, floods or ice. Mere crop failure was not included in the list of factors on which such an appeal might be based, although it was sometimes mentioned as supporting evidence in the reports of the subsequent investigations. The number of farms in each parish in Sunnfjord and Nordfjord considered to be endangered in 1723 are shown in Figures 7 and 8. The commissions clearly had strong forebodings for the future of Olden

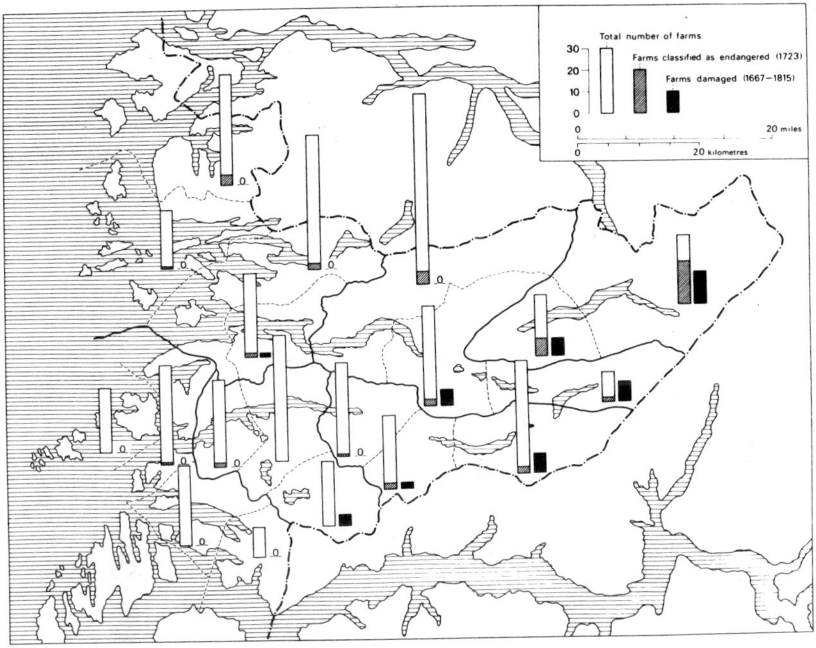

Figure 7 - Farms in Sunnfjord Fogderi considered to be endangered in 1723, compared with those actually damaged and successfully claiming reductions between 1667 and 1815.

Skipreide, where they expected the majority of the farms to be damaged, and also for Opstryn parish. In Sunnfjord they put down more than half the farms in Helgheim parish as endangered, together with a third of those in Aalhus, but they did not consider more than a handful of those elsewhere in the fogderi as likely to be claiming reductions *(avtak)*.

The commissioners' predictions may be compared with the details of actual damage which may be extracted from the voluminous records of successful avtak proceedings. The number of farms in each parish which actually suffered damage sufficiently severe as to lead to successful claims for relief is plotted on Figures 7 and 8. It emerges that the commissioners underestimated the danger in Olden and Stryn Skipreider, though not by a great deal. Damage also proved rather more marked than was expected in Vik, Holsen and Førde parishes in eastern Sunnfjord, but rather less widespread than feared in Helgheim. Trouble in Guddal had been unexpected. Much of the scatter of damage perceived as probable over the rest of the fogderi failed to materialise but the damage that did occur was serious and in some cases devastating.

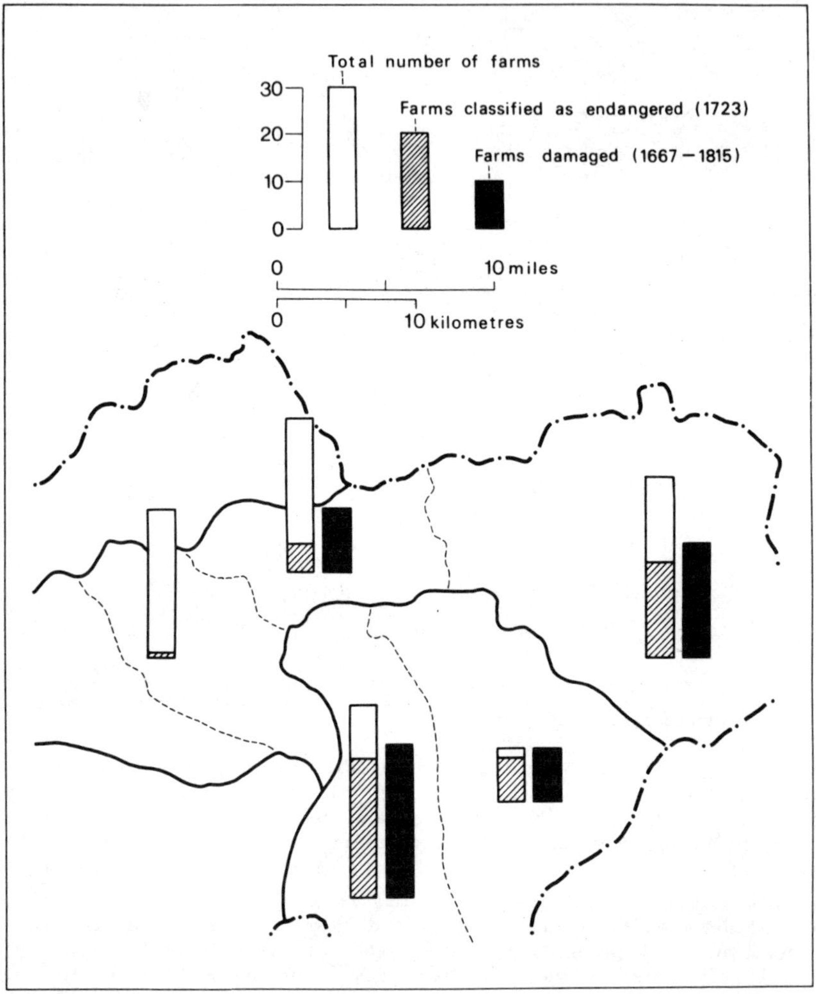

Figure 8 - Farms in Stryn and Olden Skipreider, Nordfjord Fogderi considered to be endangered in 1723, compared with those actually damaged and successfully claiming reductions between 1667 and 1815.

The proceedings that had to be employed before an avtak was granted were clearly laid down. A petition had to be made to the Provincial Governor, who issued a writ to the local Justice, charging him to hold an enquiry. The Court, formed by the Justice with six lay assessors and

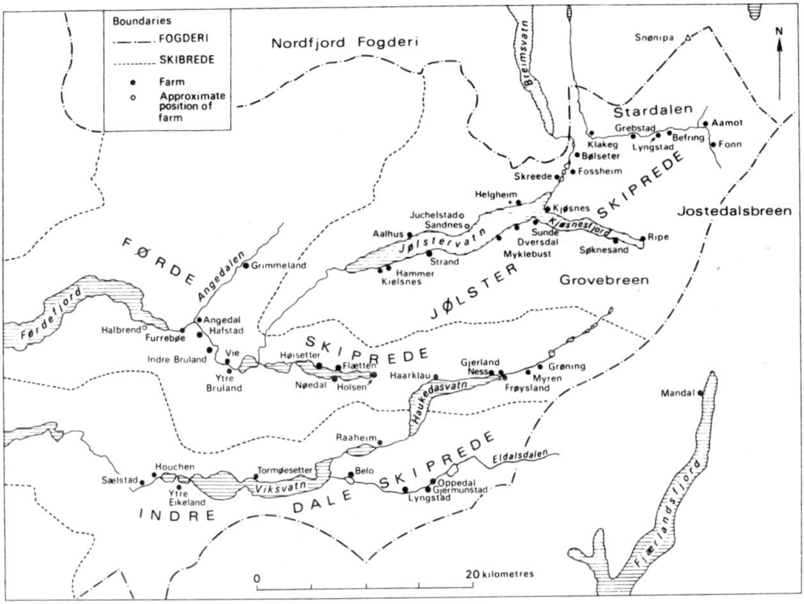

Figure 9 - Farms claiming avtak successfully between 1667 and 1815 in Sunnfjord Fogderi. (The few farms in the western part of the fogderi are named in Figure 1).

attended by the Sheriff or his deputy and any landowners involved, sat at the farm itself. All documents relating to former tax reliefs were read, and details of the farm's produce given. The Court then inspected the damage and heard evidence as to how it had taken place. The damage was then assessed separately for each bruk. A report by the Justices on the whole proceeding was read aloud at the next session of the local Ting, registered, and sent with a report on the Ting proceedings to the Provincial Governor, who forwarded it to the Tax Department of the Copenhagen Treasury. Eventually it was approved after an interval of approximately one to four years after the original incident.

The complete set of avtak records from Sunnfjord farms between 1667 and 1815 (when the tax structure was reorganised) have been examined. Sixty one Sunnfjord farms claimed relief successfully during this period, some of them more than once (Fig. 9). The resultant data have been graphed in Figure 10 in five-yearly sets. This makes it possible to compare the results with those from the Nordfjord parishes (Fig. 11). The data for this area are drawn from Aaland (1932). Aaland contents himself in many cases with giving the date for the final confirmation of reduction from Copenhagen.

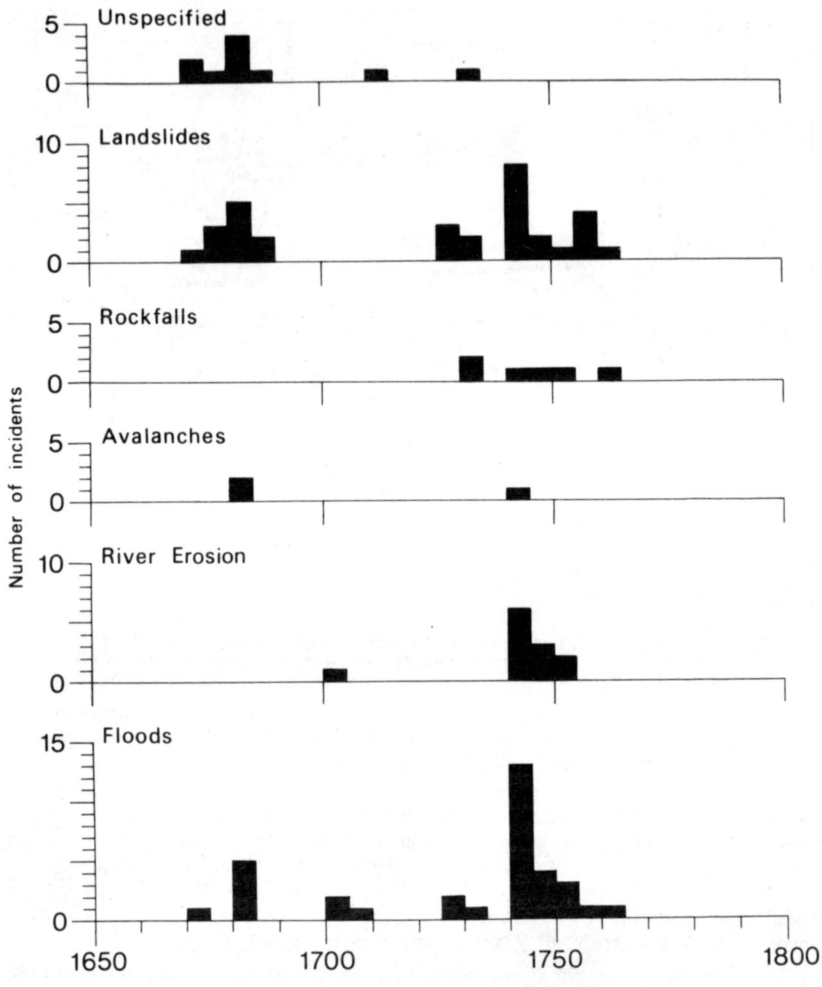

Figure 10 - Graph to show incidence of hazards causing avtak to be granted in Sunnfjord Fogderi 1667 - 1815, plotted in 5 year sets.

It emerges that environmental deterioration set in within three years of the 1667 matrikkel. Between 1670 and 1690, when the glaciers were undoubtedly swelling rapidly, farms in upper Jølster and around Jølstervatn were affected by landslides and avalanches, while farms lower down valley were subjected to floods. Thus Furrobøe, in Førde, was given a 23 per cent reduction in its assessment in 1683. Sometime before 1700, some of the coastal farms were also damaged. The exact dates are unknown as very unusually both the farmers and the authorities had lost the original documents. Thus when an inquiry was held at Vilnes in 1733, the occupiers reported that a reduction had "taken place because there was an island called Lødøen, belonging to the farm Vilnes which, because of the good sealing there, had been formerly assessed at 0.5 laup butter on the whole farm but where, neither now nor for a long time since, had there been any sealing". The assessors found both the pasture and the arable to be "very bad", as they also did at Stubseide on the northern side of Stongfjord. The reduction here "had taken place because, in times past, the farm had been so highly assessed on account of its fisheries, but the same concern was now so much in decline that they expected that they might benefit in future from the reduction made". The reports from these farms at least suggest the possibility of changes in the water characteristics off the west coast.

During the eighteenth century, incidents covering damage occurred in several groups of years and particularly affected Haukedalen and Viksdalen as well as lower Jolster. Difficult conditions were widespread with increased danger from all kinds of physical hazard. Farms damaged earlier had not recovered. In 1733 the tax authorities held an inquiry to see if the 1667 assessments could be restored; all avtak were intended to be temporary. In all, 24 farms were visited and in every case the avtak had to be confirmed.

The twenty years from 1740 to 1760 brought the greatest toll. It was during these two decades that the glaciers reached their most extended positions; a slow retreat then set in. Flooding was the most prevalent form of damage. Certain years stand out as having been particularly disastrous, notable amongst them being 1743. In December 1743 over 110 farms are known to have been damaged and claimed avtak successfully, scattered over Vestlandet from Ryfylke to Sogn og Fjordane (Øvrebø 1970). In Jølster skipreide, 12 farms made a joint petition for relief which explained that "It is, regretfully, all too well known that our lands, arable and pasture, within and without the fences, together with our buildings have been spoiled and ruined by the excessive rains of last autumn, causing flooding, landslides and rockfall". Doubtless, heavy rain in December will have caused particularly serious flooding because of associated snowmelt and enhanced runoff due to frozen ground.

No successful claims for relief were made in Sunnfjord after 1763. The percentage gaining relief was greatest in Jølster with 23 farms, with 38 per cent of the totale affected, but the highest percentage of the total assessments removed was in Indre Dale skipreide. Within Jølster, the relief granted as a percentage of the total 1667 assessment ranged from 5 per cent at Befring farm to 51 per cent at Aamot, and the average was 23 per cent of the 1667 rate. Amongst the 12 Indre Dale farms involved, the range was from 20 per cent at Gjermunstad to 75 per cent at Saelstad, and the average was 36 per cent of the 1667 assessment.

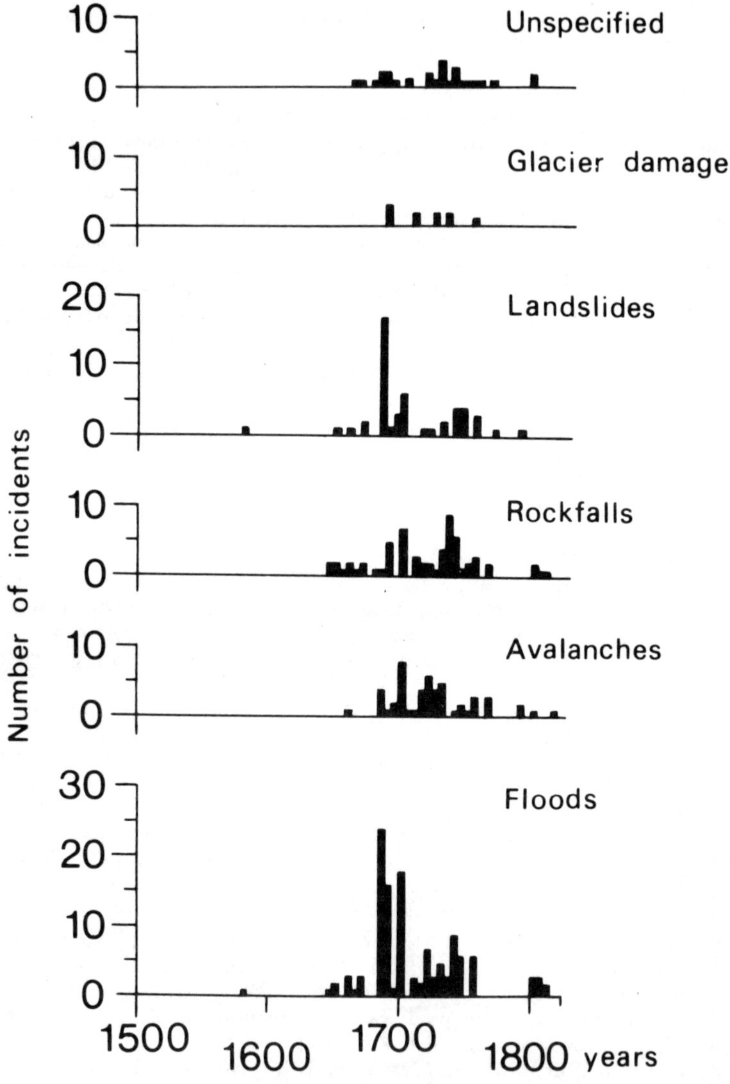

Figure 11 - Graph to show incidence of hazards causing avtak to be granted in Olden and Stryn Skipreide 1600 - 1815. Data from Aahl Aaland (1932) plotted in 5-year sets.

The relief granted on landskyld and ***skatteskyld*** (the base for tax), for the 61 Sunnfjord farms came to a total of 43 lauper, 3 punder and 10 merker and 1.5 hides. The real loss to the Treasury was much greater as the subsidiary taxes were reduced automatically and in proportion. As for the farmers, they were now able to sow only 226.5 tønner of oats as against 375.5 tønner in 1667, and to keep 858 cattle instead of the former 1387.

There is no doubt that the physical damage to farms between 1667 and 1723 was responsible for loss of pasture land which affected the number of beasts that could be kept. At Dversdal a commission in 1667 reported that

> *"we found that within the stream which is called Yttre Dversdahl inside the fenced land, two landslides had fallen from high on the mountains and across the pastures, not only causing great damage to the pastures each year, but in particular, even greater damage this year. One had fallen over a great width onto the pastures and had spread out into three branches, carrying much sand, rock and grit over arable land and pasture, as well as tearing great ditches across the fields, which appear impossible ever to clear; two other landslides which were found further out had also caused severe damage to arable and pasture though some of it, in time, might be cleared with great effort and labour."*
>
> *"Within the stream called Indre Dversdahl, we also found a gully which likewise discharges...each year and ruins the poor pasture, which moreover is nothing but scree and rocks and quite useless. Which serious damage to both Indre and Yttre Dversdahl we estimated at one whole laup butter deductions..."*

This extract from the Dversdal report is typical both from the point of view of the serious nature of the damage described and the obvious relevance to the extent and quality of the pasture. Unfortunately it is also typical in that it does not provide direct information on the relationship between the damage and the head of cattle kept. Reports from some other fogderi are more specific. Thus it is recorded at Digranes farm in Hardanger Fogderi in 1742 that "This damage has meant the loss of 10 to 12 head of cattle fodder, as well as woods, which were the farm's poor assets and which have been taken away".

Reductions were only granted in cases of serious damage. Of the 12 Jølster farms which petitioned for relief in 1743, for instance, only 5 were successful. Six of these others which had suffered more minor damage and had made it good by 1747 were not even visited by the commissioners. A number of the Sunnfjord reports make it clear that petitions were not always made even in cases of quite severe trouble. A report from Kjøsnes, on the north bank of Kjøsnesfjord, the eastern branch of Jølstervatn, describes damage by 13 separate landslides in 1756/7, which attracted relief but mentions that "as was well-known throughout the whole skipreide, a landslide several years ago had cost him 14 head of cattle, some being killed outright, some carried into the lake, and others buried beneath rocks, soils and sand". But no petition was made after this incident and no relief granted. More minor incidents or series of incidents, such as the landslides and rockfalls that took place "year after year" at Grebstad must have affected many farms which never petitioned at all, or which had their petitions turned away at an early stage, as unlikely to succeed, or finally had petitions turned down. In all, there is no doubt that serious environmental deterioration did take place in a wide zone surrounding Jostedalsbreen between 1667 and 1765.

It must be doubtful whether physical damage was adequate to explain either the total drop in the number of cattle kept in this period or its wide extent although it clearly played an important part. There are no obvious socio-economic reasons for the change. Sunnfjord farms do not seem to have been prone to innovation. As late as 1830, Kraft was writing that the agriculture here was "hindered by prejudice, poor agricultural implements and the lack of crop rotations". In our period, most of the men selected as lay assessors were unable to write. There seems no possibility that the farms had turned from cattle to some more profitable enterprise. It is possible that they may have suffered from cattle pest of some sort as they did in 1659 (Samlinger til det Norske Folks Sprog og Historie V, p.501) though no evidence of this has been found. Even so, adverse climatic conditions can enhance the likelihood of animal as well as human infections. On the available evidence it can only be concluded that a substantial fall in the numbers of cattle was probably caused by a reduction in the length of the growing season and lowered temperatures. The extension of the glaciers at this time provides completely independent evidence of the existence of these general conditions which were no doubt felt with particular severity by those farms which also suffered from the impact of mass movements and flooding. This may be exemplified by Loen farm in Nordfjord Fogderi which by 1702 had both the outlying fields and the cattle pastures ruined, the fenced-in building damaged by rockfalls and large portions of the homefields carried away by a change of river course. The fishing was ruined so "the farm does not nearly pay its way, so that the occupants, one after another, have been impoverished and have had to leave the farm, so that a number of buildings remain deserted".

ACKNOWLEDGEMENTS

Professor Andreas Holmsen has advised us over the last few years and has kindly read the text of this paper and made helpful suggestions. We also thank archivist E. Øvrebø for his advice and the staff of the Riksarchiv for their patience and helpfulness. The majority of the diagrams were drawn by Pamela Lucas, Figure 6 by Michael Young and Figure 11 by Arthur Shelley. The manuscript typing was done by Morag Meyer.

REFERENCES

Manuscript sources

Sunnfjord Tingbok A. Nr. 18 (1745-48) Fol. 233-238b.
Sunnfjord Tingbok B. Nr. 3 (1742-60) Fol. 127-127b.
Sunnfjord Tingbok A. Nr. 20 (1751-55) Fol. 198b-201b.
Sunnfjord Tingbok A. Nr. 21 (1755-60) Fol. 150-155b.

Sunnfjord Svano Birks Domkopibok (1762-72) B. Nr. 1. Fol. 24-24b.
Sunnfjord Fogderi Avtaksforetning. Reports 1667-1765.
Matrikkel over Nordfjord 1667 Vol. 44.
Matrikkel over Sunnfjord 1667 Vol. 43.
Matrikkel over Nordfjord 1723 Vol. 146. Folio 93b-139a.
Matrikkel over Sunnfjord 1723 Vol. 146. Folio 175a-326a
Martikkel over Nordfjord 1723 Vol. 147, Folio 89a-121a.
Matrikkel over Sunnfjord 1723 Vol. 147, Folio 147a-271b.
Kongelige Resolusjoner 1743 Liber F.

Published sources

Aaland, J. (1932) *Nordfjord, Innvik-Stryn-Sandane.* 2 vols., 373 and 528

Grove, J. M. (1972) the Incidence of landslides, avalanches and floods in western Norway during the Little Ice Age. *Arctic and Alpine Research*, 4, 131-138.

Holmsen, A. (1956) The Old Norwegian Peasant Community. *Scandinavian Economic History Review*, IV, 17-32.

Holmsen, A. (1958) Landowners and Tenants in Norway. *Scandinavian Economic History Review*, VI, 121-131.

Hovstad, H. (1971) Pest og Krise. *Den Norske Turistforenine Arbok*, 33-39 and 41-49.

Hoel, A. and Werenskiold, W. (1962) Glaciers and Snowfields in Norway. *Norsk Polar Institutt Skrifter*, 114, 291 pp.

Kraft, J. (1830) *Topographisk-Statistisk Beskrivelse over Kongerige Norge.* Vol. 4, 962 pp.

Øvrebø, E. (1970) Avtaksforretninger, Riksarkivet's Private Archive No. 273 Riksen. *Heimen*, XV, 83-84.

Paus, H. (1751) *Gamle Forordninger og Privilegier udgivne for Kongroget Norge.* II.2, Copenhagen, 1108 pp.

Rekstad, J. (1901) Iagttagelser fra braeer i Sogn og Nordfjord. *Norges Geologiske Undersøgelses*, 34.

Rekstad, J. (1902) Iagttagelser fra braeer i Sogn og Nordfjord. *Norges Geologiske Undersøgelses Aarbog fra 1902*, 3, 9-45.

Sandnes, J. (1971) *Ødetid og Gjenreisning.* Univeritets-forlag, Oslo-Bergen-Tromso, 385 pp.

Samlinger til det Norske Folks Sprog og Historie, V, 717 pp., Christiania, 1838.

CLIMATE OR MAN? THE EVIDENCE OF SEDIMENTS AND EARLY MAPS FOR THE AGENTS OF ENVIRONMENTAL CHANGE IN THE POST-MEDIEVAL PERIOD.

Catherine Delano Smith
University of Nottingham

Sedimentary evidence from Italy and France (Corsica) is interpreted as showing a post-medieval phase of changed environmental conditions (soil erosion, flood, siltation). Cartographic evidence is produced in corroberation of this. The argument is put forward that it is the spatial, temporal and directional coincidence of three sets of factors (physical constants, physical variables, human variables) of specific character that results in the "worsening" of conditions; the coincidence of only two sets will not have such serious consequences.

This paper is about coincidences. What has to be discussed, however, is whether it is likely that there was ever any degree of causality in these coincidences. The data is derived from long-term investigations into environmental change in Mediterranean Europe since the Neolithic period where the research emphasis is on coastal and valley changes and on the implications of these in human terms. Overt confrontation with the question of climatic change has been deliberately avoided hitherto in view of the ambiguity of so much of the 'evidence' and of the circularity of arguments derived from it but certain data implicate the climatic factor in such a way that they merit further discussion.

As far as the post-medieval period is concerned, the background is familiar enough. It has been one of major environmental change overall. Coastal plains have been created by both the natural processes of sedimentary accretion and infill and by the artificial processes of drainage and reclamation. Inland, the alluvium of flood-plains has spread more broadly across the valleys than at any other time, historic or prehistoric (Delano Smith, 1981). The record on the human side has been no less eventful. Periods of demographic growth, even population pressure and "land hunger", have been interspersed with periods of demographic decline and of retreat of the settlement frontier and the abandonment of *agri deserti*. Urban growth and industrial developments in the later Middle Ages led to increased demands for raw materials and for fuel, both of which usually had to be satisfied from the region in general if not from the local area in particular. Consequently, bare rock or secondary plant associations have replaced the forests of the hills while farmland has replaced the marshes and lagoons of the coasts and the woods of the plains. Small wonder, then, that the accepted model of environmental change in Mediterranean Europe is one of deforestation / soil erosion / accelerated and augmented run-off / flooding / siltation (Fig. 1). In this set of factors, the climatic one

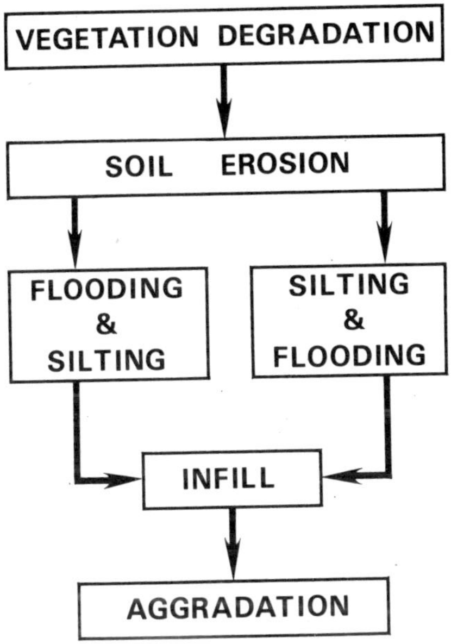

Figure 1 - Environmental change in Mediterranean Europe: a simplified model.

has no obvious place. However, the evidence of the sediments and of early maps from the areas discussed in this paper (Fig. 2) suggests there may have been more than mere chronological coincidence between maxima of late and post medieval erosion and the phase of persistent bad weather known further north in Europe as the Little Ice Age.

It is perhaps inevitable that the bulk of the evidence for the post medieval climatic deterioration (summarised by Lamb 1977, p.451-2) comes from the northern parts of our hemisphere. As far as Italy is concerned, only the Alps were directly or dramatically affected (Le Roy Ladurie, 1972). However, it is distinctly arguable whether the concomittant increase in rainfall or the sortening of the summer season would have been as beneficial in the areas further south as Lamb, admittedly tentatively, suggests (p.449). In fact, documentary evidence points the other way. In northern Apulia for instance, the exceptionally severe winters, late springs, and bad flooding of the seventeenth and eighteenth centuries took a heavy and unwelcome toll of livestock and are carefully chronicled (e.g. Dominicis, 1781). It would seem that the Mediterranean *evidence* for the post medieval climatic change is as overshadowed by the northern evidence as the disasters themselves were small, or localised, in the south in comparison with

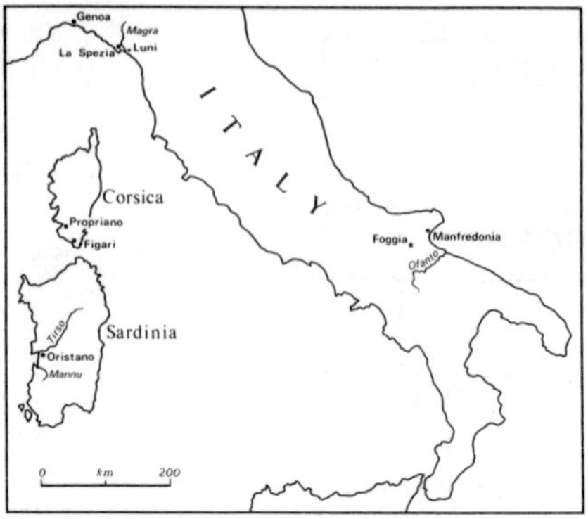

Figure 2 - Places mentioned in the text.

those of the north. The fact cannot be avoided, however, that a change in weather tendencies did affect many parts of Italy. Each piece of evidence outlined below can be interpreted as implying some sort of "worsening": increased erosion in Corsica; exceptional floods in Sardinia; current reversals along the eastern coast of mainland Italy; river course changes (carefully mapped) and repeated flooding by river and by sea in Liguria - all since the mid-sixteenth century, as far as the dating goes at present.

THE SEDIMENTARY EVIDENCE

The argument concerns both the nature of certain deposits and the agents of their deposition. The noteworthy aspect in each case is the presence of an alien element in the sedimentary profile (Fig. 3). The four profiles illustrated come from Corsica, Sardinia and Apulia. Since all have been published elsewhere (Delano Smith, 1978, 1979; Delano Smith and Morrison, 1974) it is sufficient here to summarise the relevant points. Although the object of the field investigations is to describe the changing character and outline of individual stretches of coast since Roman times, there are problems in providing precise dates. Absolute dates for individual horizons in the examples discussed are not yet available and a *terminus post quem* can be offered in only two cases.

Two of the examples come from Corsica. In both cases, the normal depositional sequence of an aggrading coast is interrupted by an alien material. This was identified as growan, the sharp-edged, gravel-like detritius that results from the erosion or decomposition of granite. The bed-rock of both areas in question is granite but since the growan appeared nowhere else in the profiles, is outstandingly coarse relative to the other deposits and, from its angularity, can be associated with neither marine nor fluvial agencies of deposition, it is regarded as

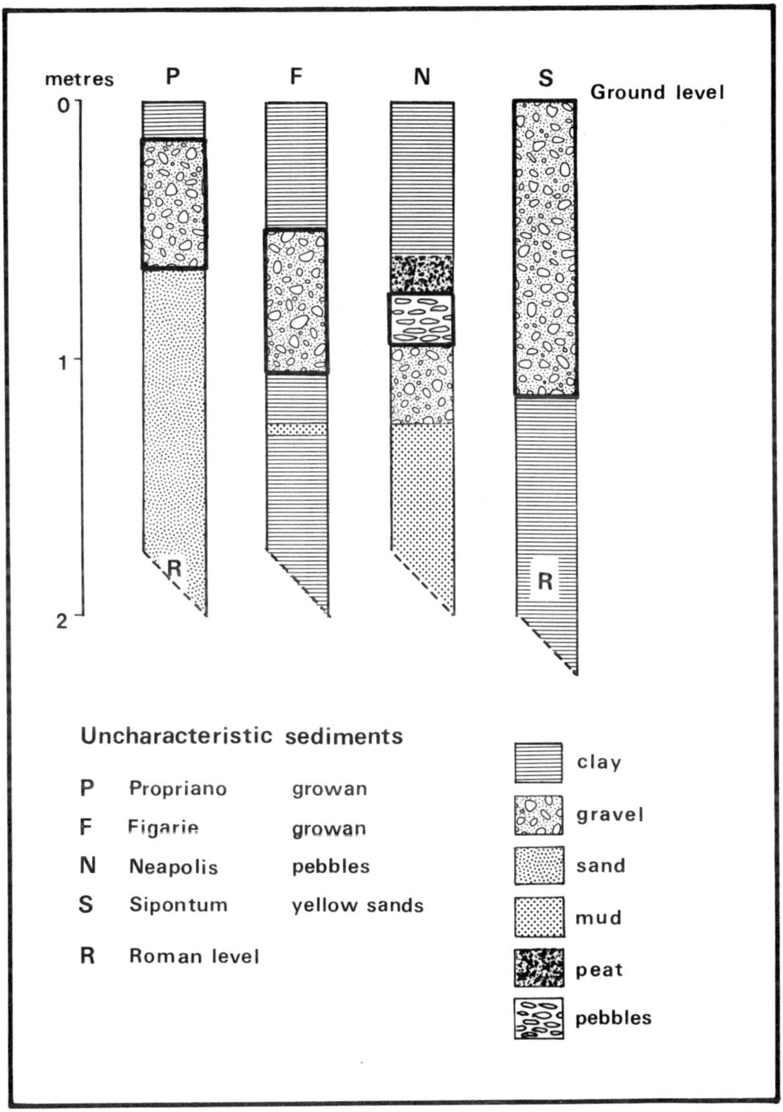

Figure 3 - The sedimentary profiles (the uncharacteristic sediments are emphasised by the heavy line).

uncharacteristic. Both at Propriano and at Figarie, the growan deposit totalled nearly 50cms. At Propriano it is covered by colluvial material which may have been artificially levelled or compacted through use of the local area for recreation. At Figarie, it is covered by nearly 50cms of the black muds and grey clays of the surviving lagoons and marshes. In both places, the growan rests on the normal coastal series: lagoonal muds and clays at Figarie, bay-head sands at Propriano. The latter are known to represent part of the infilling of the inlet that had served in Roman times as a harbour (Caesari, 1974). Although the sampling area at Figarie was chosen because of its proximity to the site of a Roman villa no artefactual evidence was found in the samples to aid dating.

The presence of growan in otherwise normal coastal sequences seems to point to something special. First, it indicates a phase of intensive erosion or weathering inland, resulting in the breakdown of the granite. Second, although its angularity shows that it has not travelled far, there must have been sufficient run-off or stream flow to have transported this material to the shore. Third, it seems that the phase was a single one, whether of short but concentrated activity or more prolonged. Fourth, as far as dating is concerned, all that can be said is that it must have been a recent event since the growan represents the penultimate deposits both at Figarie and at Propriano, that is, post-fifteenth or sixteenth century. A fifth point can be noted: that at S. Agatha, further north on the east coast of the island, unrolled sands occupy a similar stratigraphic position.

Similar conclusions were reached after field work at Neapolis in Sardinia. The western coast of Sardinia is interrupted by the gulf of Oristano (Fig. 4a). Several rivers reach the sea in this gulf, among them the Tirso in the north and, at the head of a deep inlet further south, the Mannu and associated streams. Near the head of the Mannu inlet is the site of the Punic port-city of Neapolis, occupying the edge of a rocky plateau and overlooking the now rather marshy inlet. The profile (Fig. 3) comes from the shore just below the city. Below the heavy, clayey, field soil is nearly half a metre of dark grey mud, containing much Roman building debris and potsherds swept down from the site above. This is followed by a layer rich in plant matter. Below this is a layer, about 15cms thick of pebbles, followed by one of gravels, then sandy muds and, by the time the water-table checked drilling, pure muds.

It is the pebble layer that attracts attention as the uncharacteristic element at Neapolis. The pebbles are by no means small, being up to 5-6cms across. They are distinctly flat but with very well rounded edges. Most significant of all, they are of micaceous sandstone, while the local rock is basalt. In short, the pebbles are intrusive in the sedimentary sequence at Neapolis. It is evident from their roundness that they had been transported some distance prior to deposition and the geological map confirms that they are most likely to have originated in the Tertiary strata of the low, rounded hills, long

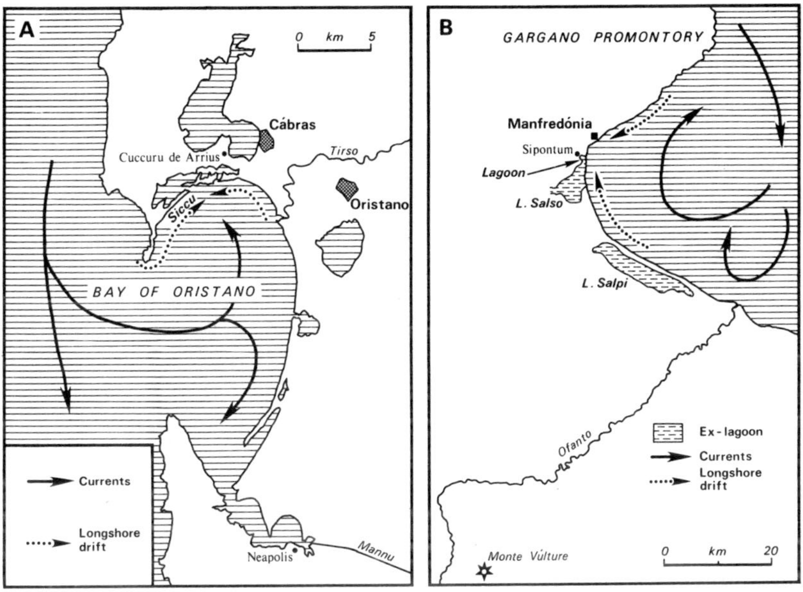

Figure 4 - Currents and longshore drift in the Gulf of Oristano (Sardinia) and the Gulf of Manfredonia (Apulia)

settled and well farmed, that border the Campidano plain on its northern side and where the Mannu has its head-streams. Again, the inference drawn is that a single, unusual, phase saw the removal and transportation of this material from the hills down to the coast over 40km away. It may be presumed that transportation was by the river Mannu in, however, a state of exceptionally high flow if not actual flood.

The final sedimentary profile to be discussed comes from the Adriatic side of the Italian mainland, the Bay of Manfredonia in northern Apulia. It is one from an extensive drilling programme in what used to be a small lagoon at Sipontum (Fig. 4b). This port-city had been founded as a colony in the Roman Republic. It flourished under the Empire and until the middle of the thirteenth century. Even at this date it was described as "a good port" in the Compasso di Navigare (Motzo, 1947) and no reference was made to problems of siltation. So the fact that, by 1250, the Hohenstaufen ruler Manfred had established the new town of Manfredonia 4km to the north must be ascribed to a preoccupation with diplomacy and with political and commercial interests in total port-control rather than to any damage sustained by the old city in the earthquakes of the previous half-century or to problems of harbour siltation.

The history of the Sipontum lagoon falls into three main stages. About the middle of the first century A.D., when an opulent, lagoonside villa was built at its southern end, the lagoon was relatively open, with freely circulating water and light-coloured, fine-textured clays and muds. Later, probably in the medieval period, conditions became brackish and marshier. By the end of the sixteenth century (see next section) the lagoon had been completely and irreversibly cut off from the sea. That this event was an abrupt one and due to exceptional weather conditions is inferred from the nature of the blocking sands (Fig. 3).

The shore of the lagoonal section of the coast of the Bay of Manfredonia is composed of sands that have been carried northwards from the mouth of the River Ofanto. These are of very dark colour, in places almost black. This colouration is due to a high heavy mineral content, derived from the volcanic rocks of Monte Vulture (Sartori and Quaratese d'Archiardi, 1966). This material can be traced along the shore of the bay as far as Manfredonia, after which the coast becomes predominantly rocky with small bay-head beaches of sands of quite different character, coarser in texture and relatively yellow in colour. At Sipontum, Ofanto sands are found throughout the sedimentary profiles and indeed they form the barrier island which created the little lagoon. However, the material blocking what used to be the main entrance to the lagoon is not this fine, dark Ofanto sand but a coarse, yellowish one, presumably of Gargano origin, containing undamaged shells of molluscs such as *cardium* and *Mactra corallina*. The significance of this is that the latter, especially, is a large and brittle subtidal bivalve which cannot have withstood much longshore movement without fragmenting.

The explanation of the temporary change of sediment source at Sipontum lies in one of the current reversals to which the Bay of Manfredonia is known to be prone (Stanley, 1972, p.336). It is well-known that strong winds can move beach-building materials in directions counter to that of the main current (as in the case of the Siccu beach in the Gulf of Oristano (Fig. 4a)). There is no problem, therefore, in accepting the presence of sands of Gargano origin at Sipontum instead of the usual Ofanto material. What is noteworthy is the scale and uniqueness of the deposit of the yellow sands. Nowhere else in the sedimentary profiles are the Gargano sands in evidence so the blocking of the lagoon was not only on such a scale as to make closure final and permanent but it appears to have been an exceptional event. The inference drawn, therefore, is that the weather conditions responsible for the arrival of Gargano sands at Sipontum were also of exceptional severity rather than duration. The dating of the closure of the Sipontum lagoon is therefore of considerable relevance to any discussion of the possible role of the Little Ice Age in environmental change.

THE CARTOGRAPHIC EVIDENCE

A search for contemporary cartographic and other archive information to complement fieldwork is still in its early stages. Nevertheless, sufficient has been discovered to show that the search is worth pursuing. While it is one matter to locate relevant documents, however, it is quite another to evaluate them fully prior to accepting the information contained in them. Most of the cartographic evidence presented here is derived from research in progress and has not yet been subjected to such an evaluation so the proffered interpretations are at best tentative. Even at face value, though, it is clear that the potential value of cartographic evidence in the study of environmental change is considerable.

(a) Sipontum, Apulia. While it was the contrasting nature of the sands at Sipontum that revealed the closure of the lagoon, the dating of this event has to rest, at present, on cartographic evidence. After a search of published regional maps of Italy, two maps were selected for their reliability and authority and for the recentness of the information they were based on, assurances derived from R. Almagia's commentary (Monumenta Cartografia Vaticana 1944-1955). The earliest is by Gastaldi and is dated 1561. The second, painted on one of the walls in the Vatican Gallery by Egnazio Danti, was executed in 1580-1. The point of concern in the context of this paper is the treatment of the coastal outline at Sipontum. Gastaldi maintained the tradition of showing the lagoon as open. He was the last authority to do so. That he was more likely to have been right than wrong is accepted on the basis of the quality of his information, known to have been first hand and not merely copied or derived from existing maps (Almagia, vol. II p.26). Likewise, it is known that Danti had sent to the provinces for new surveys and for the most recent information (ibid vol. III p.8). Again, it has to be presumed that it was according to this information that Danti showed the lagoon as closed.

Effective or completed closure of the Sipontum lagoon, which led to its eventual infilling, must have occurred in the late 1560s and in the following decade. The sedimentary evidence has shown that it was due to a reversal of the beach-building currents in the Bay of Manfredonia. While this is not an unusual occurrence in itself, the fact that the accumulation of the uncharacteristic Gargano sand in the entrance of the lagoon was on such a large scale and so permanent is interpreted as evidence of an exceptional state of weather.

(b) Liguria. Research on coastal change in the vicinity of the site of Roman Luni (province of La Spezia) also led to the Archivio di Stato at Genoa for cartographic evidence. All the maps selected for mention in this paper are in manuscript form and most of them have not been published. All require evaluation from the point of view of their information content and matters relating to this. Meanwhile, attention is drawn here to their salient points and to the possible inferences.

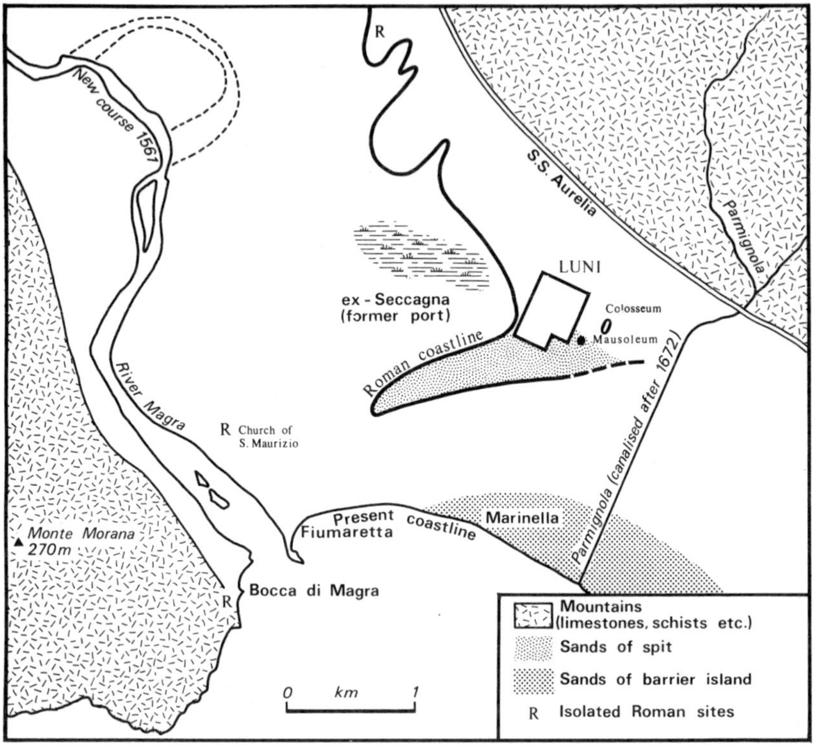

Figure 5 - Coastal outlines at Luni (La Spezia).

A number of maps concern the Luni area in particular. The site of the Roman colony is now over one and a half kilometres from the sea, separated from it by a coastal plain, much of which has only recently been reclaimed for agriculture or settlement (Fig. 5). North of Luni, the coastal plain broadens to 3km where it merges with the floodplain of the River Magra. Its evolution, therefore, has been particularly complex, involving both fluvial and marine agencies, the advance of the Magra floodplain and the build up of sand bars and barrier islands off the coast.

In 1592 Ercole Spina was also interested in these questions of coastal changes at Luni and, moreover, drew a map giving his ideas of the advancement of the Magra floodplain (kindly made available by D. Moreno, University of Genoa). He showed four positions of the coastline (Fig. 6). The date of the first is illegible because of a stain but is described on the map as *"L'anno [ ] di Nostra Salute sin qui era il porto di Luni. Talomea nel 3 "*. The last position, that of Spina's own day, is dated 1592. From where Spina derived all his information has yet to be ascertained but it is clear that the position of his shorelines imply that as much, or even more, land was built up in the years between 1500 and 1592 as during all the eight centuries between 700 and 1500 A.D.

A weakness of Spina's map is that he appears to have ignored, or was himself ignorant of, the role of the sea as well as of the Magra river in contributing to the build-up of the coastal plain above and below Luni. Our own borehole data (taken in conjunction with profiles supplied by SALT from the autostrada survey) show that it has been the action of the sea that has been in the main responsible for the manner of deposition of sediments brought down not only by the Magra but also by the small but potentially powerful streams that drain the wet, western, flanks of the Apuan Alps. By Roman times there was already a large, sandy spit lying hard against the city walls, projecting westwards and partially enclosing the broad, shallow, expanse of water that, despite its marshiness, served the Romans as a port. By the late Middle Ages this area was known as Seccagna. It is still in part a wetland, neither wholly drained nor usable. In addition to the spit, an off-shore barrier island appeared at some time opposite Luni from the south. Quite what filled the intervening space in Roman times is not yet clear (the sedimentary investigation cannot proceed without pumping) but two points have been established. One is that a period of extreme wetness, even submergence, followed the Roman period, with lagoonal clays being laid over the Roman peats in Seccagna. The second is that this period was in turn followed by one of siltation and marsh growth, despite the continuation of the marked negative isostatic tendency that affects this area and that was no doubt responsible for the submergence. The cartographic evidence illustrates this latter phase. A brightly coloured and boldly drawn map of the environs of ancient Luni has been given a date of 1626 (map B. 17 C. 936, Sarzana I; Marengo, 1931, No. 307). It shows a much reduced Seccagna, surrounded by a considerable area of marsh, together with a strip of water between the city and the barrier island on the southern side of the deserted city. Less than two centuries later, this water body had been reduced to a marsh sufficiently passable, however, to permit access to the settlement of Marinella, newly established on the barrier island itself.

To judge from the maps in the Archivio, a great deal of practical interest has had to be paid in Liguria to the state of the rivers, since a large number of them record major river course changes or are concerned with the effects of river flooding. An early example is dated 1565 (***Macchia e fiume della Macra presso Sarzana;*** Map B. 17. C.959, Sarzana 14; not in Marengo, 1931). It depicts the lowest of the Magra's meanders and records that the meander had been abandoned "four years ago" in favour of a more direct course across the meander neck. The abandoned meander is still clearly visible on the ground and on modern maps (see Fig. 5). That this was not the first time such an event had

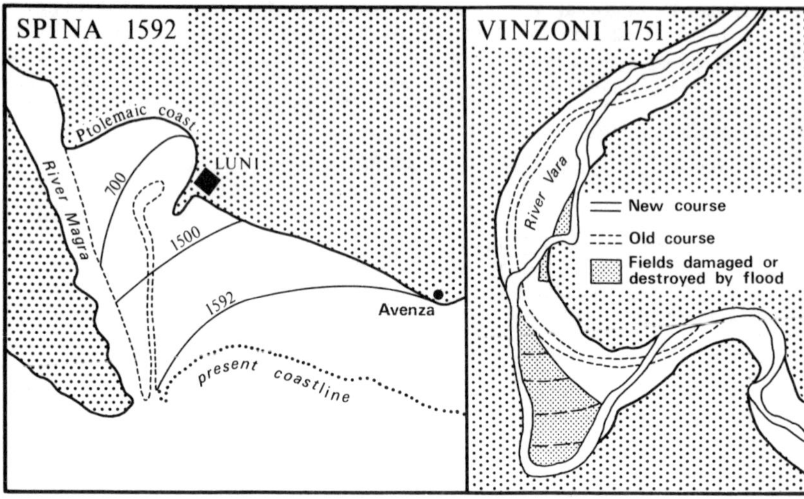

Figure 6 - Cartographic evidence for changes in coastal outline and river courses; two examples. (The position of the coastline today has been added to Spina's map).

occurred is shown by annotated lines marking a previous course or flood across the meander neck. The new (1561) course is indicated by a separate strip of paper cut to the shape of the new course and fastened into the main sheet on the right. The whole is so designed as to be able to demonstrate the old and the new situations by simply lifting or lowering the strip. Another, much plainer, map shows the area of devastation resulting by the repeated flooding of the Parmignola, southeast of the site of Luni (map B. 17. C.955, Sarzana 9; not in Marengo, 1931). It bears the clearly written date of 1672. It shows also the projected line for an artificial course for the Parmignola, probably that followed by the modern dyke, and was ordered, the title informs us, *"per divertimento e riparo delle acque della Parmignola, che devastavano li territori megliori di Ortanovo, Nicola, e Sarzana"*.

The most impressive corpus of cartographic evidence for environmental change in Liguria comes from the office if not the actual pen of the early eighteenth century surveyor Matteo Vinzoni. Vinzoni, member of an illustrious family of surveyors, was born in Levante on the 6th December 1690 and died there on the 12th August 1773. Solely from material surviving in the Archivio di Stato in Genoa, he is known to have had a hand in no less than 80 manuscript maps, of which at least 47 were his own work (Marengo, 1931, p.vii fn. 2). Either he was particularly interested in surveying natural features, as Negri suggests (1960), or much of the work that came to him was the direct result of major floods, river course changes or, in the case of beaches, storm damage. For example, at least four maps concern the same stretch of river, the meander of the Magra immediately above that already discussed. Here, in the first half of the eighteenth century, Vatican land on the west bank of the Magra has been eroded or devastated by flooding and Vinzoni's brief was evidently to record this damage and to show how the construction of palisades might prevent a repetition. One of the four maps, dated 1743 but referring in the key to floods that cut across the meander neck between 1728 and 1743, is illustrated in Marengo (Plate IX). Two others, possibly drawn as a pair, have been dated as "after 1756" by Marengo (Map B. 10 C. 626-7, Lerici 2 and 3; Marengo 1931 nos. 159-160). Another of Vinzoni's maps concerns part of the Vara, a left-bank tributary of the Magra. This was mapped in 1757 (Fig. 6), after a major course change had devastated fields and damaged a house, part of the main Ligurian highway (the Via Aurelia), a quay and a bridge (all detailed in the key). Much of Vinzoni's work was along the coast where, between 1765 and 1771, he surveyed towns and fishing villages eventually to be included in a bound set of manuscript maps presented to "il Dominio della Serenissima Republica de Genova in Terra ferma" on 2nd August 1773. One of these maps is of Final Marina. On this, Vinzoni has indicated the damage done along the shore by a previous storm (or storms). The beach had been reduced to half its former width and several buildings along it totally destroyed or damaged.

One of the problems yet to be resolved is the degree to which the cartographic record represents phases of cartographic activity rather than phases of physical change and environmental damage. Whatever the answer to this is to be, one point is already significant: the fact that from the sixteenth century onwards there were river course changes of no mean order and frequency to be mapped.

EROSION; FACTORS AND COINCIDENCES

The evidence outlined in the two sections above relates to a variety of physical changes or circumstances that are at first sight striking because they appear to have been isolated occurrences and not part of the normal local situation. Taken together, however, another aspect is significant: their chronological coincidence (Fig. 7). It remains to identify possible relationships not only between the contributary physical factors but also between these and the relevant human factors.

The interrelationships between human and physical factors in environmental change is readily appreciated in the context of river course change. The nature of a stream's course reveals much about the regime of that stream including its total annual velocity, seasonal variations in flow, and the nature of sedimentary load. The cutting off of a meander reflects a change in stream behaviour associated with an increase in the total discharge. This could be the result of any one of several factors; either an increase in run-off rates (higher up in the catchment perhaps) as a result of deforestation or of major land use

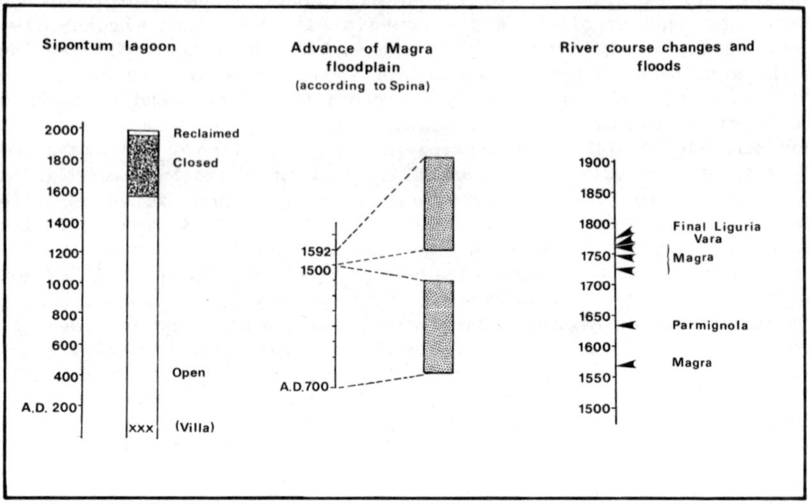

Figure 7 - Chronological summary of environmental changes evidenced in the cartographic record.

changes; or an increase in precipitation (result of a phase of 'bad' weather or climatic "change"; or an increase of both run-off and precipitation. Note that the increase in run-off - the key factor in environmental change in Mediterranean Europe since it affects soil erosion and thence flooding and siltation - results from either physical change (increased precipitation) or a change brought about by human activity (deforestation).

Not every type of area, however, will react in the same degree to such changes. Certain areas can be expected to react more positively. These are what geomorphologists class as 'high risk' in terms of erosion potential (Kirkby, 1969; Morgan, 1969). These are areas of steep slopes, predominantly impervious rocks, concentrated rainfall and, not surprisingly, a tendency towards high run-off rates unless these are firmly checked by the vegetation cover. The nature and status of vegetation in these areas is, therefore, a factor of paramount importance and any change in the vegetation cover will have a critical effect. Most at risk are climatically dry regions (with an annual rainfall of less than 600mm) (Kirkby 1969, p.235) because of the difficulty of plant regeneration without the nursery conditions of a reasonable plant cover, but the same also applies to almost any limestone area in Mediterranean Europe because of its pedological dryness. Moreover, there is no reason to limit the concept of high risk regions to physical criteria. Certain regions in Europe stand out for their vulnerability to problems of land hunger in times of population pressure or to vegetation degradation in

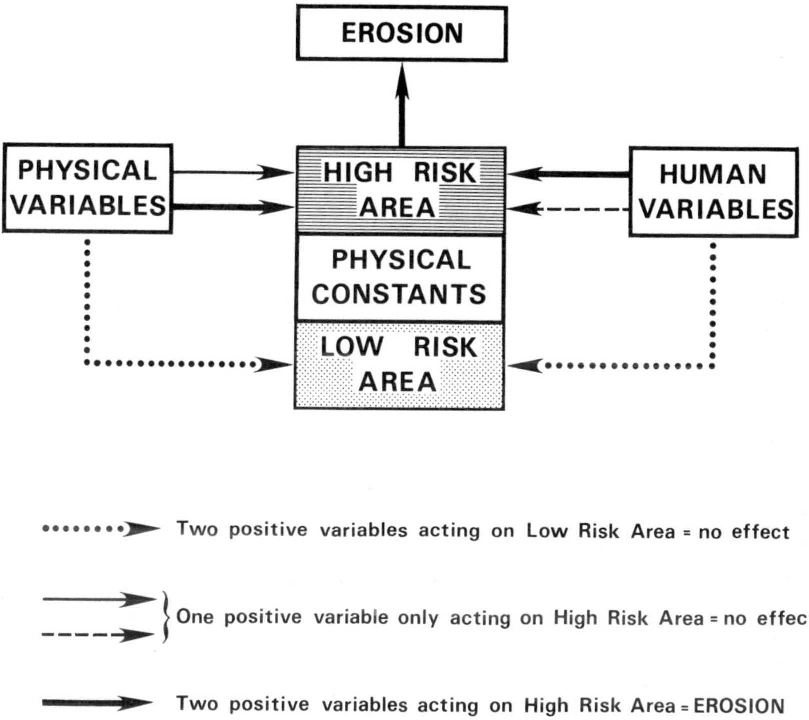

Figure 8 - Climate or Man? The effect of high risk areas of changes in land use and climate is only of consequence when these are both simultaneous and positive (tending towards erosion). Negative tendencies are not indicated.

times of industrial prosperity. Mountainous Liguria is an outstanding example of a region where there has never been sufficient room for settlement and economic activities, at least since the Roman period. Most of Corsica and much of Sardinia are also to be viewed as 'high risk' areas on both physical and human criteria. Even northern Apulia, though for quite different physical liabilities (Delano Smith, 1979 p. 276ff) and with a very different settlement and economic history, should be so designated.

An attempt to show the relationships between physical constants, physical variables and human variables in the context of environmental change is given in Figure 8. Physical constants include rock type and major relief characteristics, aspects unlikely to alter significantly in the post-Neolithic period. Vegetation cover is included since this too would not alter significantly (in the context of erosion potential) were it left alone by man. The climatic factor has to be classed as a physical variable; recent authoritative statements on the "fickleness of climate", as H. H. Lamb has expressed it, make this amply clear (Lamb, 1977; Le Roy Ladurie, 1972; Parry, 1978). Of the numerous human variables, the critical ones are those that have most direct effect on the vegetation cover: population numbers, settlement, and land use. All these factors can be qualified. The physical constants of a particular region can be classed as "high risk" in terms of erosion potential or as "no" or "low" risk. Likewise, the tendency of each of the two sets of variables can be classed as either "positive", working towards potential erosion, or "negative", working away from it. For instance, periods of demographic decline tend to result in retraction of settlement and agriculture from marginal areas and for the vegetation to be found in a progressive phase of ecological evolution.

What emerges from this view of the factors of environmental change in Mediterranean Europe is that it is the timing and effect of changes in both sets of variables, specifically in a high risk area, that is critical. A change in one variable alone is unlikely to overlap thresholds of tolerance in the physical characteristics (or constants) of the area. A worsening of climate, with increased precipitation for instance, need not result in increased run-off and erosion so long as the vegetation cover is intact, whether the region be classed as a high or low risk. Even were the vegetation to be badly degraded or removed altogether (e.g. for cultivation), still no adverse effects would necessarily result so long as the area was of the very low- or no-risk type. But the consequences of a threefold coincidence of (1) a worsening of climate in (2) a deforested area of (3) high risk class is bound to be significant.

CONCLUSIONS; THE HUMAN VARIABLES

This paper dealt first with evidence for physical changes of, given the locality, an unusual nature. The evidence has come from two quite different sources, sedimentary and cartographic. All of it, however, points in one direction: that of a post-medieval period (or periods) of considerable environmental change, arguably deleterious. It is arguable, too, whether these changes are to be ascribed directly to the climatic worsening of the so-called Little Ice Age alone or only in relation to concurrent human activity of the sort that emphasises the vulnerability of high risk regions to soil erosion and its consequences. What has not yet been put forward is evidence for variations in the intensity of human activity and for their chronological coincidence with the climatic variable.

Sufficient evidence is already available to outline the sequence of human activity in the areas in question. It also supports the contention that the coincidence is not merely academic but has led to the sort of environmental changes implied in the sedimentary and cartographic evidence. Two examples only are summarised here, concerning Liguria and Corsica.

For Liguria it has been shown (Moreno, 1973; Moreno and Maestri, 1975) that there were two phases of colonisation on the Ligurian mountains. Each was for different reasons, each had a different pattern, and each would have had, it can be argued, a different effect on the environment. In the first phase, the pressure of growing demands in the twelfth and thirteenth century (notably from Genoa) led to a substantial penetration of isolated settlement in the heavily forested mountains behind the city. This, however, was associated with increased industrial exploitation rather than with demographic pressure and the search for more land. In contrast, the second phase of colonisation, in the sixteenth and seventeenth centuries, was above all a result of land hunger at lower altitudes. In consequences, the pattern of occupation and settlement in the mountains and of the pressure on the land was different - perhaps greater, certainly far more widespread and much less selective. Wherever land (below 850m) was at all suitable for cultivation, it was cleared; nuts and roots were gathered for food; chestnuts were planted; livestock of all categories was grazed at all altitudes (whereas earlier there had been only seasonal sheep grazing above the forest line). It cannot be doubted that the pressure of human and human-induced activity first on vegetation and then on the soil would have had greater effect in terms of soil erosion in the second period of colonisation than in the first. Moreover, by this time the effects of a worsening of weather would have been felt all too directly on the bared soil and cleared mountain sides. Hardly surprising, therefore, if the later period of colonisation is found to coincide, as it appears to, with the increased incidence of river course changes and flood and storm damage.

The same applied to Corsica, where the land hunger of the eighteenth and nineteenth centuries has been well documented (Perry, 1967) and ecological changes since prehistoric times have also been studied (Reille, 1975). It is interesting to find that much of the blame for the disappearance of the old oak forests in the eighteenth century A.D. is ascribed to the excessive number of free-ranging pigs that illegally wandered the mountains at this time, providing the destitute but demographically increasing peasantry with an important element of their subsistence diet. Settlements expanded and multiplied and terraces, now mostly abandoned and heavily recolonised by vegetation, were built as high as possible up mountain sides to provide scraps of arable land. If the product of heavy erosion of the newly cleared land, the growan found in the sedimentary profiles at Figarie and Propriano, could be dated accurately, it would come as no surprise if the dates were precisely from this period. Certainly, in view of the stratigraphic position of the growan layers, such a date (seventeenth-eighteenth century) seems highly reasonable.

Such arguments are hardly conclusive, it has to be admitted. In part this reflects the incompleteness of the research but it also reflects a fundamental weakness in the study of climatic change as a factor in historical geography at local or regional scales: the impossibility of *proving* causal relationships in the absence of detailed, instrumental, climatic records. Rather than conclude on a wholly negative note, however, it is worth pointing out that in every case it is the local sequence of human events and the physical conditions of the local area that counts. The precise combinations of regional characteristics apply only there. Demographic trends and land use histories can be highly diverse. What may happen in one district at one time does not necessarily happen in another till later, if ever. The diverse regions of Mediterranean Europe might well repay close attention by students of climatic change, for it is evident that the consequences of the Little Ice Age in Mediterranean Europe were negligible or severe according to the particular circumstances, physical and human, of each region at a particular time.

ACKNOWLEDGEMENTS

Much of the fieldwork referred to in this paper was carried out with financial aid from The British Academy, The Society of Antiquaries and the Sopintendente della Antichita (Genoa) or during the year I held the Susette Taylor Travelling Fellowship at Lady Margaret Hall, Oxford. I also owe thanks for help in the field to numerous people, in particular: M. Raskolnikoff (C.N.R.S. Paris), Professor J. and Mrs. L. Jehasse (University of Clermont Ferrand and Alèria, Çorsica), C. Puxeddu (Mogoro, Sardinia), Professor S. and Mrs. F. Tinè (University of Genoa). Mrs. E. Pyper carried out the sediment analyses in my own Department; Christopher Lewis drew the maps; Tracey Allen typed the manuscript; to all three I am grateful.

REFERENCES

Casalis, G. (1833 ff) *Dizionario Geografico - Storico - Statistico commerciale degli stati di S.M. il Re Di Sardegna.* 25 volumes.

Caesari, J. (1974) *Rapport à M. le Directeur des antiquités historiques de la Corse sur les sondages archéologiques éffectuès à Propriano, Corse* (unpublished).

Delano Smith, C. (1976) Tell-tale Snails. *The Geographical Magazine*, Vol. XXX, 210-1.

(1978) *Daunia Vetus: terra, vita e mutamenti sulle coste del Tavoliere.* Amministrazione Provinciale di Capitanata, Foggia.

(1979) *Western Mediterranean Europe: a historical geography of Italy, Spain and Southern France since the Neolithic.* Academic Press, London and New York.

(1981) Valley changes: some observations from recent field and archive work in Italy. In G. Barker and R. Hodges (eds.) *Archaeology and Italian Society: Prehistoric, Roman and Medieval Studies. Papers in Italian Archaeology II.* British Archaeological Reports (International Series), Oxford.

Delano Smith, C. and Morrison, I. A. (1974) The buried lagoon and lost port of Sipontum (Foggia, Italy). *Nautical Archaeology*, 3.2, 415-28.

Dominicis, F. N. (1781) *Lo Stato Politico ed Economico della Dogana della Dogana della Mena della Pecore in Puglia.* Napoli, 3 volumes.

Kirkby, M. J. (1969) Infiltration, Throughflow and Overland Flow, and Erosion by Water on Hillslopes. In R. J. Chorley (ed.) *Water, Earth and Man*, Methuen, London 215-227 and 229-238.

Lamb H. H. (1977) *Climate: Present, Past and Future*, Volume 2, *Climatic History and the Future.* Methuen, London.

Le Roy Ladurie, E. (1972) *Times of Feast, Times of Famine; a history of climate since the year 1000.* Allen and Unwin, London.

Marengo, E. (1931) *Carte Topografiche e corografiche manoscritte della Liguria e della immediate adjacenze conservate nel R. Archivio di stato di Genova.* Ministero del Interno, Genoa.

Moreno, D. (1973) La colonizzazione dei Boschi d'Ovada nei secoli XVI-XVII. *Quaderni Storici*, 24, III, 977-1016.

Moreno, D. and Maestri, S. de (1975) Casa rurale e cultura materiale nella colonizzazione dell'Appennino genovese tra XVI e XVII secolo. In *I Paesaggi Rurali Europei*, Atti del Convegno Internazionale, Perugia 1973, 389-407.

Morgan, M. A. (1969) Overland Flow and Man. In R. J. Chorley (ed.) *Water, Earth and Man*, Methuen, London, 239-255.

Motzo, R. (1947) *Il Compasso di Navigare.* Cagliari.

Negri, T. O. de (1960) Matteo Vinzoni e la corografia della Liguria nel settecento. *Annali di Ricerche e Studi di Geografia*, Anno XVI, No. 1. 1-37.

Perry, D. T. (1967) Economy, landscape and society in La Castagniccia (Corsica) since the late eighteenth century. *Transactions, Institute of British Geographers*, 41, 209-221.

Reille, M. (1975) *Contribution pollenanalytique à l'histoire tardiglaciaire et holocène de la végètation de la montagne Corse.* Thèse d'Etat, University of Aix - Marseille, 2 volumes.

Sartori, F. and Quaratesi d'Archiardi, L. (1966) Studio mineralogico di una sabbia della foce dell'Ofanto. *Atti della Societa Toscana di Scienze Naturali*, Memoria A. 73, 112-132.

Spano, B. and Pinna, M. (1956) *La Spiagge della Sardegna.* Centro di Studi per la Geografia Fisica, Universita di Bologna, Bologna.

Stanley, D. J. (ed.) (1972) *The Mediterranean Sea. A Natural Sedimentation Laboratory.* Dowden, Hutchinson and Ross, Stroudsbury

Tennant, R. (1885) *Sardinia and its Resources.* Stanford, London and Spithover, Rome.

The Mediterranean Pilot (1880 ff) Vols. 11 and III London, HMSO (Hydrographic Department).

Ward-Perkins, B. (1978, 1979, 1980) *Luni Interim Reports.*

THE DESERT MYTH EVALUATED IN THE CONTEXT OF CLIMATIC CHANGE

Merlin P. Lawson
Department of Geography,
University of Nebraska (Lincoln)

Charles Stockton
Lab of Tree-Ring Research
University of Arizona

Students of the American frontier have assumed that the myth of the Great American Desert was derived from the notions of a few men rather than from the probable reality of the experienced environmental conditions. In a sense the reality of the desert has not concerned western historians. Incredibly, the "illusion" of the Great American Desert has not been considered from a climatological perspective, i.e. no serious attempts have been made using historical methods and techniques to establish the actual climate of the region during the period of formation of the desert image. Explanations concerned with the derivation of the notion of a trans-Missouri desert as perceived by early American explorers neglect the possibility of climatic variability sufficient enough to justify the disparaging accounts of the territory. By establishing the statistical relationship between tree growth and climatic characteristics, spatial reconstructions of moisture conditions in the western United States permit analysis of exploratoy experience in perspective with climatic stress. One must conclude that Stephen Long, the explorer responsible for naming the Great American Desert in 1820, witnessed serious drought conditions on the High Plains exceeding that experienced by previous expeditions, thus suggesting that the desert notion had substantial merit. Two geographies of the trans-Missouri region existed in the first half of the nineteenth century, those of conception (cognition) and scientific objective reality. To differentiate one from the other, we must continue to reconstruct the climate of the past.

"The theory of a great American desert stretching over boundless wastes in the interior of the continent has been one of the most persistent ideas in the historical development of our nation. Based upon the meager facts obtainable by indirect methods, this theory has been, largely, the product of the vivid imagination of writers..." (Blackmar, 1906, 101).

"And so, in history and vision, the Great American Desert was created, but it has gradually disappeared from the maps, and likewise from the minds of the people, as they slowly realized the facts of settlement." (Blackmar, 1906, 102).

Historical analysis of the desert myth as an *idée fixe* that erroneously originated within the imaginations of travellers, explorers, philosophers and historians emphasized the view that human nature has played a dominant role during the formation of this geographical notion. Attempts to characterize geographical imagery of the trans-Missouri basin have proven most enticing to students of the frontier West, to the extent that a complex historiography of the desert image has evolved (Bowden, 1976).

There exist two "geographies" of the Great American Desert: those of conception (cognition) and of scientific objective reality. Unfortunately, adequate balance has not been extended toward the reconstruction of past geographic realities of the interior West (Lawson, 1971). Over a half century has passed since John K. Wright (1925) cautioned, "The impress of *geographic conditions* themselves upon the nature of geographical thought is an alluring subject which has not been much investigated" (italics mine). The evolution of the geographic idea of the "Great American Desert" has been influenced not only by human and cultural factors but by *geographic conditions* contemporaneous to exploration. To know the reality of the plains environment would provide a standard for assessing the discrepancy between the real world and the images conveyed by the many individuals describing the region; it is to know if these notions are based on *geographic conditions* or *geographies of the mind* (Lawson, 1976).

In a sense the reality of the desert has concerned neither western historians nor historical geographers. Neglected has been the possibility of a corresponding moisture deficit of sufficient intensity to justify disparaging exploratory accounts of the territory. Overlooked has been an admonition expressed by Isaiah Bowman as early as 1935, emphasizing that a climatic boundary (particularly in the plains) represents a mean position of a transition belt rather than a rigidly placed line. "All our climatic belts expand and contract with yearly variation", Bowman contended. "In individual years, a third of the United States may be 'desert'". It was once the fashion to laugh at the extension of the phrase 'Great American Desert' printed in large block letters upon our maps across the High Plains" (Fig.1). The drought of the 1930's suggests "that in Long's day (1819-20) and in Fremont's day (1842-1844) it may not have been so foolish" (Bowman, 1935,56).

Incorporated in Bowman's challenge was an appeal that investigators use indicators such as lake level fluctuations and tree ring analyses to study climatic changes in marginal areas often stricken with drought. But until recently, historical geographers have not attempted environmental assessment of the Great American Desert with the aid of tree-ring data. One such attempt was that of G. Malcolm Lewis who sought to determine the reasons for the "emergence of Gilpin's concept of the Great Plains" (Lewis, 1966). A few years later, Lawson (1971, 1976) extensively discussed the applications and limitations of dendroclimatological reconstructions and their possible utilization for interpreting the climatic landscape during cis-Rocky Mountain exploration. This initial study demonstrated the unique potential for reconstructing moisture conditions from tree-ring chronologies but was hampered by a lack of temporal and spatial specificity. Climatic interpretations were generalized for cumulative ten year periods prior to particular expeditions rather than emphasizing yearly situations. Dendroclimatological indices were derived from only eight locations, in most instances peripheral to the interior grasslands. Finally, because three different species were used to develop indices of growth, climatic interpretation became complicated by variant growth-response functions.

In the intervening years since Lawson's initial attempt to interpret exploratory climatic conditions, the accumulation of additional

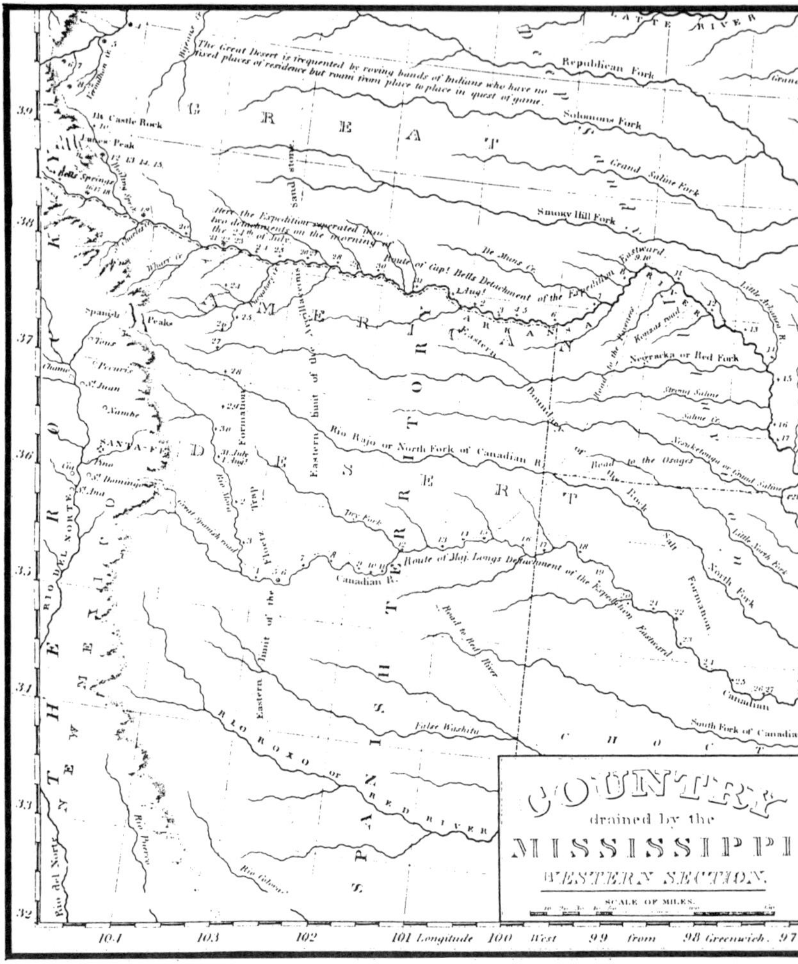

Figure 1 - Part of Major Stephen H. Long's map showing the 'Great American Desert' within the "Country Drained By The Mississippi, Western Section," 1823.

collection sites and the application of multivariate analytical techniques, provides renewed potential for circumventing previous limitations (Fritts, 1976). Thus, the purpose of this paper is to derive a greater understanding of the climatological landscape experienced during the exploratory process within the Great American

Desert as revealed by recent innovations in the derivation of climatic information from ring-width variance.

The opportunity for extracting climatic information from tree-growth indices increases as the forest environment becomes more limiting to plant growth (Fritts, 1974). The subhumid conditions of the interior plains provide variable periods of water stress (below-normal moisture) that produce growth responses of narrower-than-normal annual growth-rings. The units of ring widths can be calibrated wih climatological variables by constructing a statistical model (Fritts, 1966). Such relationships can then be applied to reconstruct climate from tree-rings for earlier periods lacking a climatic record (La Marche, 1974; Fritts, 1976).

For over a decade, the National Weather Service has employed the Palmer Drought Severity Index as an indicator of persistent moisture conditions (Palmer, 1965). The Palmer Index was designed to evaluate the scope, severity, and frequency of prolonged periods of abnormally wet or dry weather. Since it integrates the effects of the past weather over weeks and even months, the Palmer Index provides the best evaluation of the overall soil moisture situation (Table 1).

TABLE 1

Drought Classification by Palmer Drought Severity Index

Palmer Index (PDSI)	Degree of Drought
> +4.0	Extremely wet
+3.0 - +4.0	Severely wet
+2.0 - +3.0	Moderately wet
+1.0 - +2.0	Mildly wet
-1.0 - +1.0	Near normal
-2.0 - -1.0	Mildly dry
-3.0 - -2.0	Moderately dry
-4.0 - -3.0	Severely dry
< -4.0	Extremely dry

The autoregressive nature of the Index permits the representation of cumulative moisture conditions throughout numerous months by a given value. Thus, it is plausible to seek a relationship between an annual increment of tree growth and a particular monthly Palmer Index value (Stockton and Meko, 1975). Because tree growth is commonly complete by the end of July, this month was chosen as representative of moisture conditions for comparison with annual growth indices.

In general, the method of reconstructing a drought history from tree-ring indices and available PDSI records is to derive prediction equations from the period for which both sets of data exist, and then to use the prediction equations to assign values of PDSI to the earlier period for which only the tree-ring collection sites in the western United States were used in the reconstruction procedures. To simplify calculations, the 204 climatological divisions within the study were combined to form 40 regions (Fig. 2).

Figure 2 - The study area. Solid lines delineate regional boundaries, the approximate geographical centres of which are marked by black dots; regions are identified by number; triangles mark tree-ring sites used in the initial reconstruction; asterisks mark recently collected tree-ring sites not included in the initial reconstruction.

Although the situation is much improved as is evident from the depicted data, there remain large areas of the Great Plains unrepresented by tree-ring data. Thus, it was necessary to use a reconstruction technique that yielded spatial patterns of PDSI from spatial patterns of tree-ring indices. Consequently, Great Plains moisture conditions have been reconstructed indirectly, from patterns of tree growth in surrounding areas.

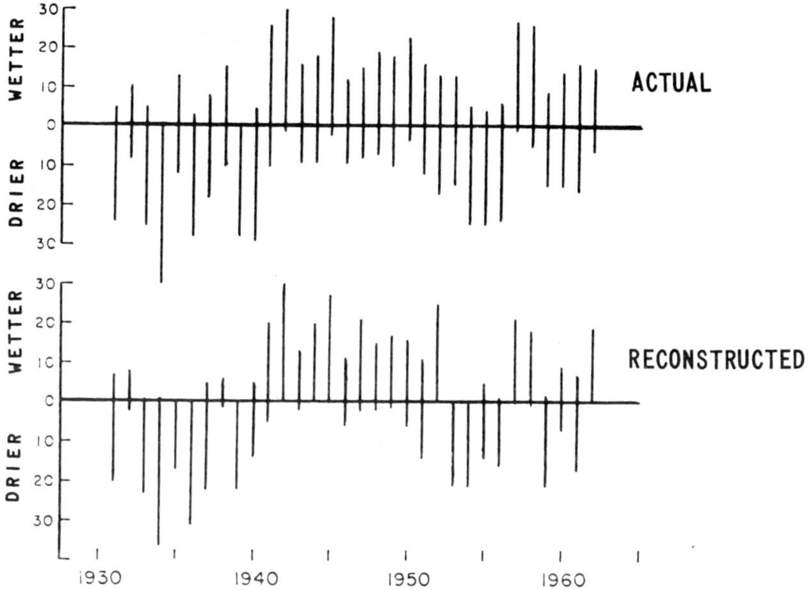

Figure 3 - Actual and reconstructed drought-area index for calibration period. Vertical lines give the number of regions wetter than normal and drier than normal, where a normal PDSI lies between ±1.0.

Mathematical equations to produce the reconstructions were derived in four steps:

1) Eigenvectors of PDSI were extracted from the 1931-70 spatial patterns of July PDSI. This procedure defined the various orthogonal patterns that accounted for most of the mathematical variance in PDSI.
2) Similarly, eigenvectors were extracted from the 1700-1963 spatial patterns of tree-ring indices.
3) Canonical analysis was applied to the amplitudes of the important PDSI and tree-ring eigenvectors for the period of overlap between two data sets (1931-1963). This yielded a matrix of coefficients from which amplitudes of PDSI eigenvectors could be constructed from amplitudes of tree-ring eigenvectors.
4) This matrix was applied to the 1700-1930 tree-ring data, and corresponding values of the July PDSI for each of the 40 regions were obtained.

Good agreement of the predicted values with the actual values in the calibration period is necessary for valid reconstructionn of past conditions. To check the accuracy of the reconstructions for the period 1931-1962, the time series for the predicted and actual data were analysed. In general, the agreement was good (Fig. 3), with the predictions tending to be conservative (more regions reconstructed to have been normal than actually were). Independent verification of the tree-ring reconstructions prior to the calibration period were also conducted. Palmer Index values were computed for numerous weather stations with records extending back into the nineteenth century. The reconstructed values continued to reflect favourably the actual conditions. As an additional approach to verifying the reconstructions, independent tree-ring records, not included in the calibration grid, were employed to assess qualitatively the validity of the retrodicted data.

The resulting reconstruction of PDSI can be generalized as a time series (1700-1962) depicting the number of regions with values above +1.0 (mildly wet to extremely wet) as compared wih those divisions having indices below -1.0 (mildly dry to extremely dry) (Fig. 4). The long term record appears to be positively skewed, suggesting that the period of calibration (1931-1962) was relatively drier than the over-all record. In terms of both spatial extent and relative magnitude, drought conditions west of the Mississippi River were most severe during the 1920s. Apparently, this was a dramatic reversal to the excessively wet regime which prevailed during the previous three decades.

Thus the climatic variance during the first half of the twentieth century exceeded both the wet and dry extremes of the previous two hundred years. Perhaps this knowledge adds perspective to Bowden's contention (1976) that the fluctuating Plains environment of this period led to discernable changes in the historiography of the desert notion. Evidently, prominent historians of the Great American Desert such as Turner (1894), Webb (1931) and Malin (1961) witnessed greater variability in the Plains climate than the subjects of their writings, the American Pioneer.

If the historiography of the regional interpretation was affected by geographic conditions of the 20th century should we not examine the extent to which the climate landscape could have influenced the exploratory imagery formed by those who witnessed the character of the region? To achieve this goal, reconstructed moisture conditions have been mapped in conjunction with the particular route followed by initial American expeditions into the trans-Missouri West.

Although controversy continues as to the origins of the notion of a trans-Missouri desert, most would agree that the first American expedition seriously to assail the potential of the region for future settlement was Zebulon Pike, who traversed the Louisiana Territory in 1806 and 1807 (Lewis, 1965; Dillon, 1967; Bowden, 1969; Alford, 1969). Extreme drought persisted at this time in the Great Basin with moderately dry conditions extending eastward across the southern plains to the Gulf coast (Fig. 5). The upper Louisiana Territory including most of the Missouri Basin had experienced above normal moisture since the

Figure 4 - Long-term reconstructed drought-area index for drier than normal (Palmer Index less than -1.0), and wetter than normal (Palmer Index greater than +1.0), A.D. 1700 to A.D. 1962.

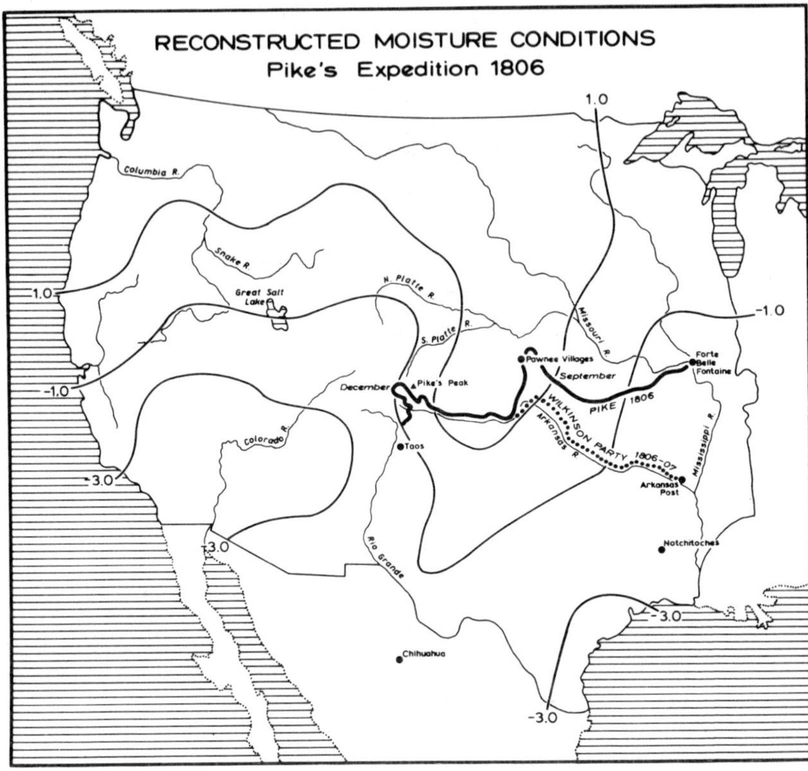

Figure 5 - Reconstructed moisture conditions for 1806 relative to Pike's exploratory route. Isolines indicate Palmer's Moisture Index.

turn of the 19th century. Pike's route across the plains to the Rocky Mountains appears to have been within this transitional zone. There is no reason to believe that exceptionally dry conditions prevailed prior to departure from Fort Belle Fontane on July 15, 1806. Meteorological observations en route provide evidence that the expedition was not hampered by any extended absence of rain (Pike, 1895). Early in the spring of 1807, after suffering greatly with winter cold, Pike and his men were intercepted by a Spanish patrol, detained at Santa Fe and ultimately taken under arrest as suspected spies to Chihuahua, Mexico. By July of this year as Pike returned to the states, the entire prairie-plains region was dominated by above normal moisture (Fig. 6). Thus, the possibility of climatologically anomalous factors must be rejected as satisfactoy explanation for Pike's critical and disparaging assessment of the Louisiana Territory. This evidence provides

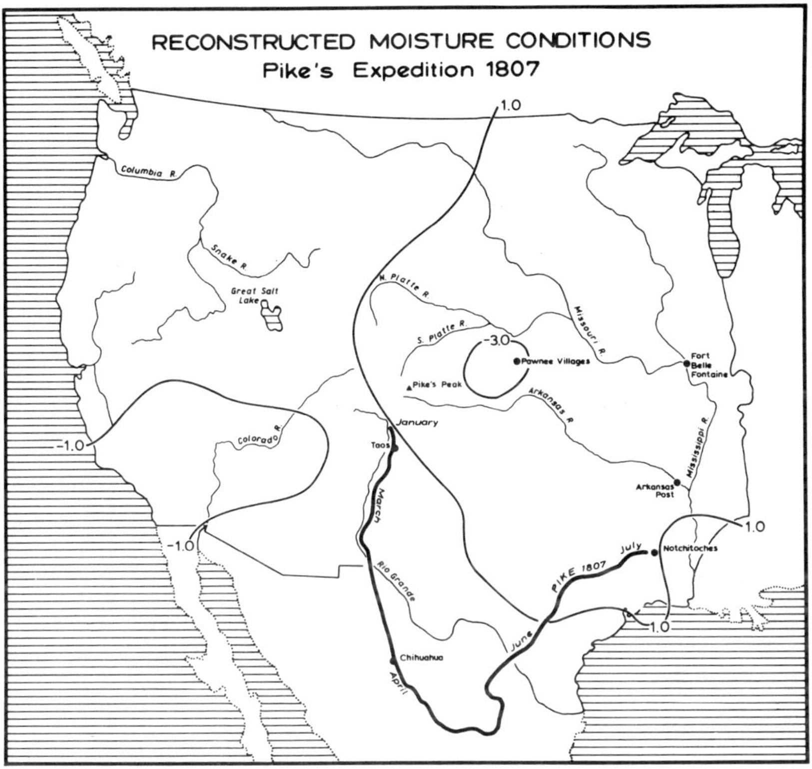

Figure 6 - Reconstructed moisture conditions for 1807 relative to Pike's exploratory route.

additional import to the continued efforts of historical geographers to analyze the behavioristic aspects associated with Pike's evaluation of the interior of Louisiana (Lawson, 1972).

An abnormally moist regime continued to prevail in the Missouri Basin the remainder of the decade culminating with severely wet conditions by 1811. During this year two unofficial expeditions crossed the Big Muddy. Accompanying these "excursions" were an adventuresome young Lawyer, Henry M. Brackenridge and John Bradbury, an English botanist representing the Botanical Society at Liverpool. Published in their respective journals are accounts of frequent showers and even severe storms which provide verification for soggy conditions indicated by the reconstructed indices. Considering that their travels through upper Louisiana corresponded with one of the most extensive wet periods since 1700 it is not surprising that Bradbury's assessment of the regional climate omits any reference to marginal or insufficient moisture. Bradbury concludes, "it will not only be peopled and cultivated, but it will be one of the most beautiful countries in the world."

And so it remained for Major Stephen H. Long to convince both the government and its people that a barrier existed in the form of a "Great Desert" to prevent an over-extension of the nation's populace westward. Edwin James' *Account* of the scientific expedition from Engineer's Cantonment up the Platte to the Rocky Mountains and return through the Arkansas and Canadian River valleys provided a vivid daily description of the Plains region as a Great American Desert, "uninhabitable by a people depending on agriculture for their subsistence". Accompanying the report was a separate atlas including a map of the entire Mississippi basin prepared by Long and his topographical assistant Lieutenant William H. Swift. The map was to provide new information concerning the physiography between the forbidding Rockies on the west and the diminutive Appalachians to the east; but particularly significant was Long's designation of a "Great Desert", in block letters extending from the South Platte to the Canadian River. The Long expedition had cartographically delimited the idea of a Great American Desert, an idea later promoted by the nation's romantic historians but assailed as a geographic misconception by the Plains historians of the early twentieth century. The former hypothesized the existence of an authentic desert east of the Rockies as "a consequence of exclusive reliance on secondary and promotional sources", while the latter minimized the environmental deficiences of the Plains during what we know now to have been the most humid period in over two and a half centuries (Nowden, 1976, 125).

Major Long had been instructed by the Secretary of War, John C. Calhoun, to acquire a thorough and accurate knowledge of the country. Did his eastern origins impugn his ability to render an objective and scientific assessment of the natural resources of the prairie-plains province?

Examination of the dendroclimatic reconstructions of 1820 appears to vindicate Long's scientific expedition. Three years of abnormally low moisture culminated in extreme drought throughout the lower portion of the western interior (Fig. 7). Long's route passed through the core of this moisture deficiency. In relative terms, the region identified by Long as the Great American Desert was experiencing drought stress exceeding that of the 1930s. Only one other period, that of the late 1720s, surpassed the intensity or extent of the drought witnessed by Major Stephen H. Long. When one considers the magnitude of this moisture anomally, augmented by pressures of overgrazing by buffalo, perhaps it is less reasonable to refer to the desert as an illusion (Watson, 1969).

For Long and his scientific cadre, the "impress of geographic

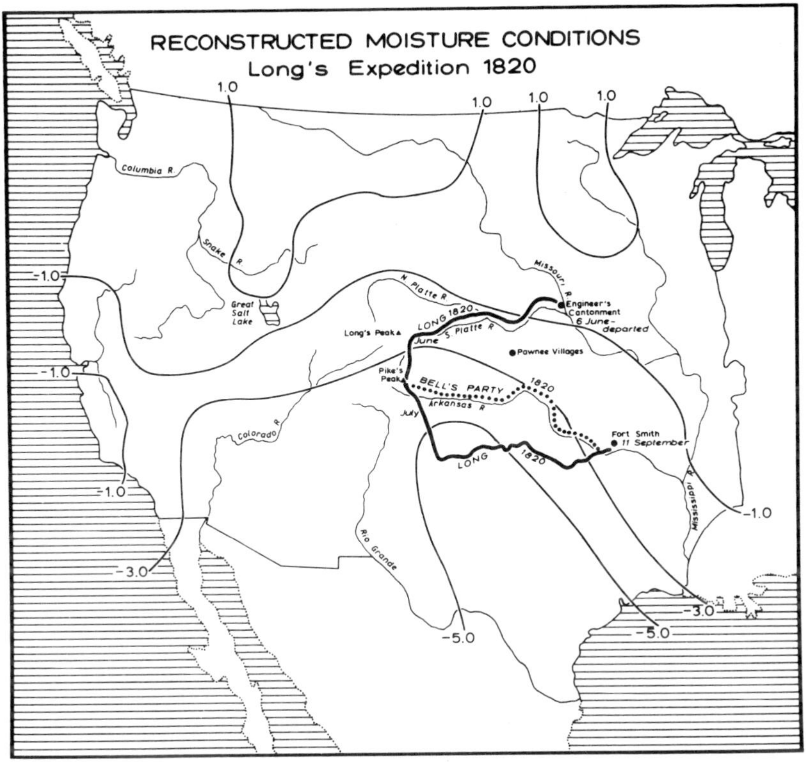

Figure 7 - Reconstructed moisture conditions for 1820 relative to Long's expedition.

conditions" represented a reality never since experienced in this cis-Rocky Mountain west.

Scholars concerned with the explorations and settlement of the trans-Mississippi West have recently focussed much of their effort upon identifying the regional myths and environmental misconceptions that developed during the "exploratory process". The problem with their studies is that the standard against which the congitive errors are measured is assumed but not known, that is, the "real" past environment is taken to be unchanging. The climatic reconstructions made in this study demonstrate this working assumption to be untrue, erroneous and further demonstrate the possibility of reconctructing the climate of large portions of the plains region during incipient exploration and settlement.

REFERENCES

Alford, T.L. (1969) The West as a Desert in American thought Prior to Long's 1819-1820 Expedition. *Journal of the West*, 8 (4), 515-525.

Bowden, M.J. (1969) The perception of the Western Interior of the United States, 1800-1870: A Problem in Historical Geography. *Proceedings of the Association of American Geographers*, 1, 16-21.

Bowman, I. (1935) Our Expanding and Contracting Desert. *Geographical Review*, 25, 43-61.

Dillon, R.H. (1967) Stephen Long's Great American Desert. *Proceedings of the American Philosophical Society*, 111, 93-108.

Fritts, H.C. (1966) Growth-rings of trees: their correlation with climate. *Science*, 154, 973-979.

Fritts, H.C. (1974) Relationships of ring widths in arid-site conifers to variations in monthly temperature and precipitation. *Ecological Monographs*, 44 (4), 411-440.

Fritts, H.C. (1976) *Tree Rings and Climate*. Academic Press: New York.

La Marche, V. C. Jr. (1974) Paleoclimatic inferences from long tree-ring records. *Science*, 183 (4129), 1043-1048.

Lawson, M.P. (1971) A Dendroclimatological Interpretation of the Great American Desert. *Proceedings of the Association of American Geographers*, 3, 109-114.

Lawson, M.P. (1972) A Behavioristic Interpretation of Pike's Geographical Knowledge in the Interior of Louisiana. *Great Plains-Rocky Mountain Geographical Journal*, 1, 58-64.

Lawson, M.P. (1976) *The Climate of the Great American Desert*. University of Nebraska Press: Lincoln, 135 pp.

Lewis, G.M. (1965) Early American Exploration and the cis-Rocky Mountain Desert, 1803-1823. *Great Plains Journal*, 5, 1-11.

Lewis, G.M. (1966) William Gilpin and the Concept of the Great Plains Region. *Annals, Association of American Geographers*, 56, 33-51.

Malin, C. (1961) *The Grasslands of North America: Prolegomena to its History with Addenda*. Lawrence, Kansas: By the author.

Palmer, W.C. (1965) Meteorological Drought. *U.S. Weather Bureau Research*, Paper 45. U. S. Department of Commerce Washington, D.C.

Pike, Z. (1895) *The Expedition of Zebulon Montgomery Pike*. Edited by Elliot Coues, New York 3 volumes.

Smith, H.N. (1947) Rain Follows the Plow: The Notion of Increased Rainfall for the Great Plains, 1844-1880. *Huntington Library Quarterly*, 10, 169-193.

Stockton, C.W. and D.M. Meko (1975) A Long-term history of drought occurrence in Western United Statess as Inferred from Tree Rings. *Weatherwise*, 28, 244-249.

Turner, F.J. (1894) The Significance of the Frontier in American History. *American Historical Association Annual Report for 1893*. Washington, D.C.: Government Printing Office.

Webb, W.P. (1931) *The Great Plains*. New York: Grosset & Dunlap.

HISTORICAL SOURCE MATERIALS AND THE RECONSTRUCTION OF PAST CLIMATES

N. Huckstep
Climatic Research Unit,
University of East Anglia

The paper discusses the importance of rigorous analysis of source materials in the reconstruction of past climates, and notes that much previous work in this field has not carried out such an analysis. A brief account of the work in progress in this field at the Climatic Research Unit is also given.

In order to reconstruct the climatic conditions of the past, many disciplines must be brought together and caused to work together. History is one such discipline, furnishing information about past climate from written sources. The analysis of these sources and the reconstruction of past climatic conditions is the business of the historical climatologist.

It would seem to be axiomatic that historical source material needs to be tested and evaluated as rigorously as any other sort of data, yet in the short history of historical climatology, this has rarely been done. Many data which have been previously accepted as reliable are not. The present writer is aware of his distinguished predecessors in the field. Not least among these is Professor Hubert Lamb, whose signal contribution to Historical Climatology none can afford to ignore, and whose efforts to provide funds for the continuation of the work he began have benefited many.

The last century has seen an increasing interest in past climatic variations, and it is during this time that historical climatology has been established as a discipline. On a similar time scale, the concerns which occupy historians have become more various, extending beyond the traditional fields of politics and great men's characters to include the environment, sociology and anthropology. While not wishing to engage in academic empire building, I hope to show in this paper how it is important that those working in historical climatology have a sound background in historical methods. I shall also give an account of one particular method of climatic reconstruction which has been used at the Climatic Research Unit and with which I have been closely involved. This work has been carried on in a collective fashion, and I should like to mention the contributions made to it by Graham Farmer, Martin Ingram, Astrid Ogilvie and Tom Wigley as well as that of Professor Lamb, referred to above.

To date, much of the work done in Historical Climatology has relied for its input of climatic data upon works known as compilations. These are compendia of citations from various historical sources which contain weather information. The genre has a long, if largely undistinguished, history. Wendy Bell and Astrid Ogilvie (1978) have discussed with admirable clarity the difficulties inherent in the use of these works as source materials for the reconstruction of past climates. For the purposes of this paper it will suffice to highlight the main reasons for the unreliable nature of these compilations and to look at one or two examples of the genre in detail.

The unreliability of most compilations is chiefly due to the fact that they lack any conception of how historical source materials should be treated. I list below the principle questions which should be asked of a source in order to determine its reliability. The ideal source which tells us that it rained very hard at Bruges at midday on the third of March 1607 will have been written at Bruges on that date by a reliable author and will be available either in a contemporary manuscript or in an authoritative printed edition. Where weather information is concerned such first class source material is the exception rather than the rule, so the answers to the questions listed below will be of the first importance:

When was the source written?
By whom was it written?
Where was it written?
Why was it written?
If the source is printed, is it available in a good edition?
If the source is in manuscript, is the manuscript original and if not, is the textual transmission of the manuscript known?

In the case of the rain shower at Bruges mentioned above our (hypothetical) ideal source would give answers as follows:

March, 3rd, 1607.
By a diarist or chronicler whose other work is known to be reliable.
Bruges
As part of a weather diary.
The source is available in an original manuscript which can be consulted with ease.

As I have said, above, such ideal sources are the exception rather than the rule. If one gives the above answers a value of +10, one will naturally prefer the hypothetical Bruges diarist to a source which produces the next set of answers, which would have a relative value of 0.

1849
By a Belgian historian in a work on the history of the wool industry.
Brussels.
To prove that the wet spring of 1607 adversely affected the wool industry.
The source is available in a first edition, but there is no scholarly apparatus and the author does not indicate the source of his weather information.

Both of the above examples are extreme but the weather compilations contain many more citations from the second sort of source than from the

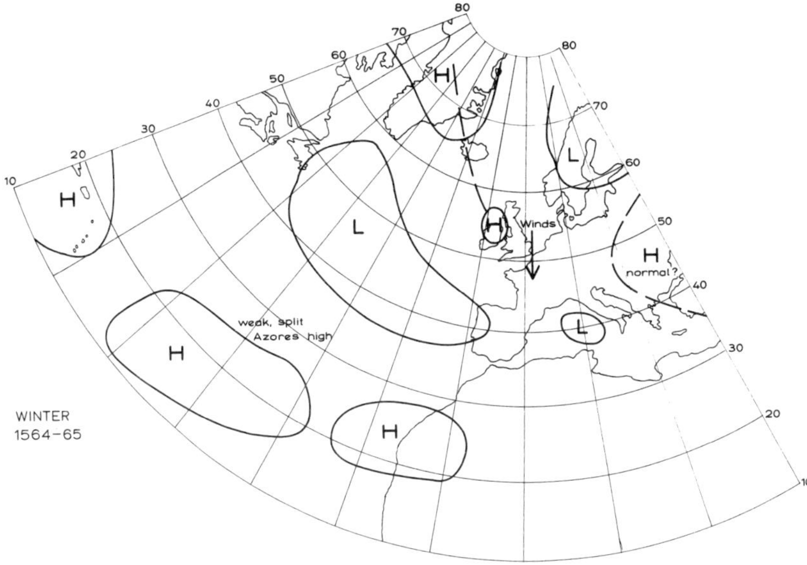

Figure 1 - An atmospheric circulation map for December, January, and February 1564-5 based on historical sources.

first. The genre must, therefore, be used only with the utmost caution. It occurs to me that it may to some, seem strange to reject the nineteenth century source. There are good reasons for doing so: it is temporally remote from the event described; the author may not have been primarily concerned with the recording of exact weather information; and, since standards of source criticism (i.e. the objective evaluation of the reliability of historical source material) and of textual integrity were generally rather low in the nineteenth century, where no information about sources is included in later works which reproduce them, it is safer to reject the information.

Historians would refer to the first example of a source as a primary source, while the second would be described as secondary. The works of Buchinsky (1957), Van der Linden (1924) and Weikinn (1958-61) rely heavily on secondary sources and these are compilations on which much of the previous work in historical climatology has been based.

In the latest work on climatic reconstruction done at the Climatic Research Unit, the more rigorous methods of source analysis described above have been used. One cannot deny that these methods end by being qualitative rather than quantitative: much is left to the historians' judgement. Qualitatively, the work which produced the two circulation maps in Figures 1 and 2 is an improvement on what has gone before.

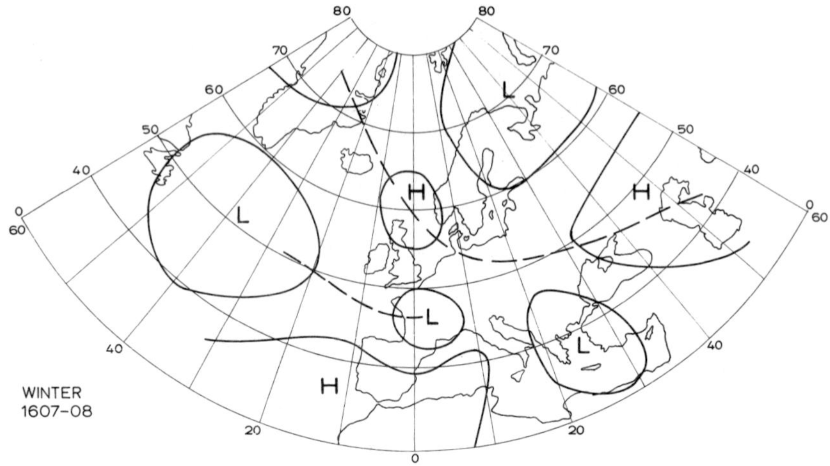

Figure 2 - An atmospheric circulation map for December, January, and February 1607-8 based on historical data.

Anyone working in the field of historical climatology must admit that we still know very little and that much work, practical and theoretical needs to be done before we can even begin to think of ourselves as workers in an exact science. It is in this spirit that I offer the following account of the work on climatic reconstruction recently carried out at the Climatic Research Unit.

For some time now the historians in the Climatic Research Unit have been engaged in building up a data bank of historical weather information for Europe. We have tried to include in this data bank only those sources which score high on the scale of 0 to 10 referred to earlier. We have more information for the British Isles than for the rest of Europe because the local archives of these islands are rich in primary source materials, mostly in manuscript form, which contain weather information. The same is almost certainly true of the archives of other European countries but neither the time or funds have been available to permit visits to these depositories.

Using this data bank, which is continually increasing in size and scope, we have now begun our attempts at the reconstruction of past climates. So far, two approaches to this problem have been considered. Firstly, a subjective method which relies on the combined skills of a group of climatologists and secondly, a more objective method based on statistical techniques such as eigenvector analysis. The circulation maps which form figures one and two are the result of the first method. Work is still in progress on the development and use of the second.

In the first method, the historians furnish a group of climatologists with a map on which are plotted, in an abbreviated form, the items of weather information extracted from reliable historical sources. They then independantly draw the circulation maps which they think will accurately represent the historical information with which they have been furnished. Their individual efforts are then combined, and maps such as those in figures one and two are the result.

In this paper I hope to have shown that rigorous data testing is essential to any attempt at climatic reconstruction which is based on historical sources. The work in this field which is being carried out at the Climatic Research Unit is not complete, but we hope that a considerable amount of the data collected as well as other research will be published in order to promote discussion and cooperation in this field.

REFERENCES

Bell, W. T. and Ogilvie, A. E. J. (1978) Weather Compilations as a source of data for the reconstruction of European climate during the mediaeval period. *Climatic Change*, I, 333-348.

Buchinsky, I. (1957) *O Klimate Proschlovo Russkoi Ravin.* Gidrometeorologiskoi Izdatyestvo Leningrad.

Weiknin, C. (1958-61) *Quellentexte zur Witterungsgeschichte Europas von der Zeitwende bis zum Jahre 1850.* Akademische Verlag, Berlin (D.D.R.), 3 vols.

Van der Linden, E. (1924) *Chronique des événements météorologiques en Belgique jusqu'en 1834.* Académie Royale de Belgique, Brussels.

CLIMATIC CHANGE AND ETHNIC MOVEMENTS IN THE EARLY IRON AGE

William C. Brice
Department of Geography
University of Manchester

In the twelfth and eleventh centuries B.C. the countries bordering the Aegean and Eastern Mediterranean were overrun by invaders from the North, with profound historic consequences. The intruders appear to have been pastoralists driven south by the onset of cooler conditions in Central Europe.

The end of the great Bronze Age civilisations of the Aegean and the Levant was widespread, simultaneous, abrupt, and complete. In mainland Greece, in Crete, in Anatolia, and the East Mediterranean seaboard, bureaucracies ceased to function, cities collapsed, literacy ended, and a Dark Age of four centuries began. Only in the island of Cyprus was there some continuity, where refugees from the Peloponnese preserved something of their art, their dialect and their writing. Egypt managed to withstand the storm when in 1186 B.C. Rameses III turned back the "Land and Sea Raiders" at the edge of the Nile delta. In the well-known reliefs of Medinet Habu we see these intruders engaged in sea-combat, and at the same time moving by land in solid-wheeled carts with wickerwork super-structures (Desborough, 1972). These are not vehicles of war, but general-purpose rustic carts as in present-day Anatolia. The passengers are not warriors, but a mixed population, including children. The "combined operations" force in fact was acccompanied by a retinue of families and camp-followers, and we seem here to have to do with something more like a tribal migration than an organised military campaign. From the Egyptian records we know the names of some of these tribes, among whom were evidently the eponymous ancestors of the later populations of Palestine (the Peleset), Sardinia (the Shardana) and Cilicia (the Danuna - cf. Adana, Edgerton and Wilson, 1936, 53). Evidently, when rebuffed from Egypt, they dispersed to the North and West and settled wherever they could find room.

From the evidence of dialects, tradition and archaeology there was much movement in Greece, as the Doric dialects drifted southwards into the Morea and Crete, and the North-West speech-forms intruded on their heels into Epirus (Myres, 1930). Ionic crossed the Aegean eastwards from Attica, while the Arcadian forms, displaced doubtless by the intrusive Doric, crossed the sea to Cyprus. At some time in this disturbed period between 1200 and 800 B.C. the Etruscans took ship from Western Anatolia to Western Italy, and the "Terremare" cultures entered the Po Valley from the Alpine passes. All in all, the archaeological and linguistic record accords with the impression left by the Odyssey, or the Aeneid, of an era of adventurers, pirates and freebooters who took advantage of the general disturbances following the intrusions of the tribes, the collapse of administration and commerce, and the enforced shift of refugee communities.

After the collapse of the Mycenaean palaces, people took to the hills, and the typical Iron-Age town was a hill-site like Lato in Crete: steep and defensible and some way from the coast and the risk of pirate raids (Pendlebury, 1939, 329ff, plate VII/I). Other innovations of the Iron Age were the cheap iron rapier, in place of the more costly leaf-shaped bronze slashing sword, and the small round hoplite targe which replaced the figure-of-eight Mycenaean body-shield. The shaped and tailored garments of the Bronze-Age Aegean were replaced by the draped and slung wraps fastened by that ubiquitous and characteristic appurtenance of the Early Iron Age, the fibula or safety-pin. At the same time the elaborate stirrup-jars of Bronze-Age times, with their free-field paintings, were replaced by simpler shapes whose derivation from wooden prototypes is confirmed by their simple ornament of tectonic patterns of bands and concentric circles which were clearly inspired by carpentry styles.

We note further that the characteristic architectural element of the Bronze-Age palace of the Aegean had been the cluster of mud-brick flat-roofed rooms around a central court. If these structures were raised to two or three storeys, the court became a light-well and the rooms were approached by open colonnaded stairways. In the Iron Age, by contrast, the typical house was a single-storey hall with colonnaded porch, pitched roof and central hearth, whose prototype would have been the wooden house of the Balkans and Central Europe.

The most detailed representations of the wooden megaron in the early Eastern Mediterranean are the rock-cut tomb-facades of Lycia. Although these date from rather later in the first millennium than the Dark Age, the style was doubtless traditional. Some of these facades show an upward recurve on the supporting beams at ground-level, and this feature can be matched in a mobile dairy of beams and withies, now in the ethnological museum in Sofia, which was dragged on its sledge-runners round the Carpathian summer pastures by the Vlach shepherds.

In short, the evidence of archaeology, ethnology and linguistics is cumulative that the intruders who entered the Eastern Mediterranean at the onset of the Iron Age, with such devastating effects on the established civilisations, were nomadic or semi-nomadic tribes-folk, accustomed to making use of wood for buildings, utensils, boats and carts, and originating in the mountainous regions of the Northern Balkans or Central Europe.

In 1965, Rhys Carpenter applied himself to the question of the cause of the onset of the Dark Age in the Mediterranean. In his J. H. Gray Lectures for 1965 (published as ***Discontinuity in Greek Civilization***, C.U.P. 1966) Rhys Carpenter began with an 1850-year cycle of droughts hypothesised by H. H. Willett (1953). From this, he went on to presume

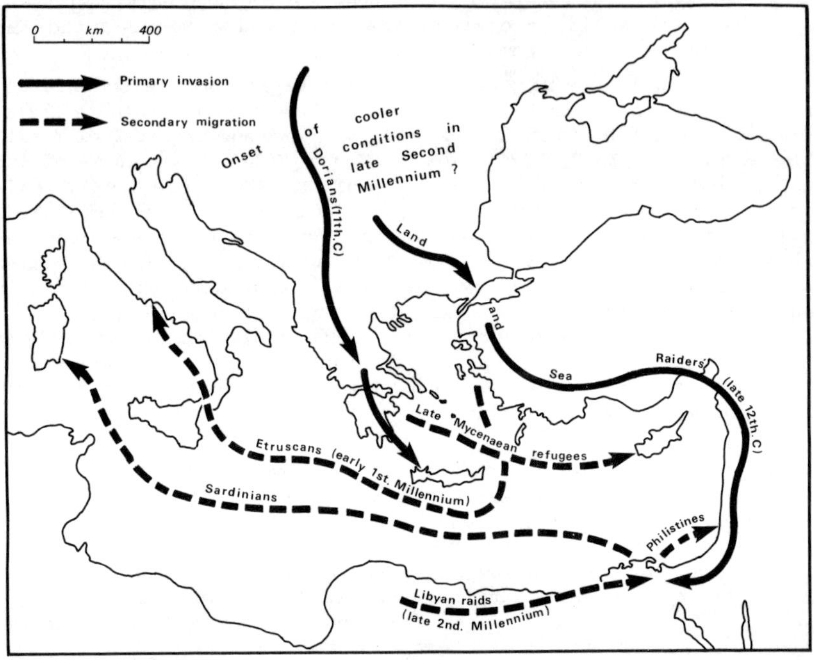

Figure 1 - Suggested lines of migration to the Mediterranean from the north, and their repercussions, in the late second and early first millenia B.C.

a prolonged drought or series of dry years in the Mediterranean starting at the end of the twelfth century B.C. and continuing for some four centuries. The ensuing damage to agriculture and the economy generally would have so weakened the established civilisations as to leave them vulnerable to attack. The next crisis of this kind in the 1850-year cycle preceded Slav invasions of the Balkans in the seventh century A.D.

The weakness of this theory is that agriculture in the Mediterranean world has always depended heavily on irrigation and is well equipped to cope with a shortage of rainfall. In any case, it is unlikely that drought would have afflicted simultaneously regions so far apart as Northern Greece, Anatolia, and the Levantine coast, which have somewhat differing climatic regimes. Further, a series of dry years might be expected to be followed by a return to normal and some reversion of conditions in considerably less than the four centuries of the Dark Age with which we have to do. Above all, famines do not normally result in the instant evacuation or collapse of whole cities.

It is suggested, rather, that if a climatic event were involved it would have been more generalized and widespread, more long-lasting, and of such a nature as to drive the tribes southwards rather to tempt them into a vacuum where, had there been a massive drought, there would have been little left to attract them.

Now we know that the general rise in sea-level, which continued throughout the period of the post-glacial "Climatic Optimum", ceased during the second millennium B.C. as a consequence of the gradual cooling of the climate of Europe through the Sub-Boreal epoch (Fairbridge, 1961). This cooling culminated in the markedly cold and damp conditions of the Sub-Atlantic phase in the last five-hundred or so years B.C.

The critical season for the transhumant communities of the Balkans and Western Asia is the summer. Winter is a time of bare survival in caves or barns but, as soon as the snow-melt permits, the flocks are taken over the mountain passes to the higher pastures where they are fattened. Every day on the summer alps is important and if, through a gradual cooling of the climate, the season of high grazing should grow shorter, the herdsmen will be tempted to look southwards for mountain pastures that are accessible for a longer season.

It is suggested, therefore, that just such a movement took place, as a consequence of a general cooling of the climate in the Balkans at the onset of the Iron Age. Why the critical break-through should have happened when it did is a further question. It could be that a drought in the Mediterranean weakened resistance and let in the intruders, as Carpenter argued.

Certainly the strong fortifications at the late Mycenaean mainland citadels of Tiryns, Mycenae and Gla are evidence of some threat, internal or external, before the final collapse. The speed with which pastoral tribes, accustomed as they are to a para-military system of authority, can convert themselves into an organized fighting force when they meet opposition to their movements is manifest in the history of the Arab conquests of the seventh century and the Mongol raids of the twelfth century A.D. However, the menace to the Mycenaean palaces would be not only from direct attack but also from insidious sabotage of commerce and water and food supplies by herdsmen overrunning the territory of states dependent on the complex bureaucratic control of labour, rations and trade.

There is an analogy in the invasion of Anatolia by the Seljuk Turks in the eleventh century of our era (Brice, 1955). The impulse which drove them out of Central Asia could have been a shortage of grass following the rise of temperature in middle and high latitudes which marked the "lesser climatic optimum" towards the end of the first millennium A.D. There was pressure on the Eastern Byzantine frontier in the Taurus and Anti-Taurus mountains both before and after A.D. 1071; but the crucial collapse of resistance happened in that year at the battle of Manzikert, after which in a brief span of a few years most of

Anatolia was "nomadised" and changed from Greek speech to Turkish, from Christianity to Islam, and from urban life to rural. There was much intermarriage and movement of population, and a fundamental change in styles of architecture, craftsmanship, epigraphy, and of course belief and ritual.

The "nomadisation" of the Levant in the earlier period involved also a breakdown of city life. In architecture, economy and epigraphy the Aegean World of the Bronze Age belonged with the older civilisations of Egypt, Mesopotamia and Syria. Quite suddenly towards the end of the second millennium its aspect changes, and it comes within the orbit of the mountain and forest cultures of Central Europe. It is suggested that this shift was the consequence of the slow onset of cooler and damper conditions which culminated in the Sub-Atlantic epoch. The shortening summers of the Carpathians forced the tribes to move southwards, and the practice of seasonal nomadism or transhumance became established in the Mediterranean region, with devastating effect on the Bronze Age palaces.

REFERENCES

Brice, W. C. (1955) The Turkish Colonisation of Anatolia. *Bulletin of the John Rylands Library*, 38, 18-44.

Desborough, V. R. d'A. (1972) *The Greek Dark Ages*. Benn, London.

Edgerton, W. F. and Wilson, J. A. (1936) *Historical Records of Rameses III. The Texts of Medinet Habu*. University of Chicago Press, Chicago.

Fairbridge, R. W. (1961) Eustatic changes in sea-level. In L. H. Ahrens et al. (eds) *Physics and Chemistry of the Earth*, Pergamon, London, 4, 99-185

Myres, J. L. (1930) *Who Were the Greeks?* University of California Press, Berkeley.

Pendlebury, J. D. S. (1939) *The Archaeology of Crete*. Methuen, London

Willett, H. W. (1953) Atmospheric and oceanic circulation as factors in Glacial-Interglacial changes of climate. In H. Shapley (ed) *Climatic Change*, Harvard University Press, Cambridge (Mass), 51-71.

THE STRATIGRAPHY AND PALYNOLOGY OF RECENT OMBROTROPHIC PEAT A SOURCE FOR THE RECONSTRUCTION OF AGRICULTURAL AND CLIMATIC CHANGE

Keith Barber
Department of Geography,
University of Southampton

Linked pollen and plant macrofossil diagrams from raised peat bogs can provide evidence for concurrent climatic and vegetational changes. Analyses from Bolton Fell Moss, Cumbria, show a close correspondence between peat stratigraphy and climatic change, but there is no discernible effect of climate on agriculture and the amount of cleared land.

As a historical biogeographer I welcome this opportunity to alert colleagues to the potential record of both climatic and land-use change which exists in the uppermost metre or so of peat in ombrotrophic bogs - those which are raised above the level of influence of mineral ground water and which receive all their moisture from the atmosphere. Such bogs are present in the lowlands of Britain as true raised bogs - discrete entities of peatland with marginal streams and a morphology not unlike an upturned plate - and in the highland areas certain parts of the aptly-named blanket bogs are also ombrotrophic. As such they may react to changes in the precipitation / evaporation ratio by changing their peat-forming plant communities in certain characteristic ways, and by changes in the humification of the peat. Where sections of stratigraphy exist in such bogs, due to drainage, peat-cutting and erosion, such changes may be picked out in the field and characterised in detail by various laboratory methods of analysis. The peat also, of course, contains pollen, and closely-spaced pollen counts from the upper horizons, in conjunction with macrofossil and stratigraphic analyses, can give a record of ecological and climatological changes over the last two millenia or more. An account of the methods used and their applicability will be found in Barber (1976).

The author has conducted such analyses in detail from a bog site at 110m O.D. in northern Cumbria, Bolton Fell Moss (grid reference NY 4969), as part of a research project on the problem of peat-bog growth mechanisms. What follows here is an outline of the results which bear on land-use and climatic changes in this border area from pre-Roman times onwards - full details may be had from my Ph.D. thesis (1978), which is to be published in early 1981.

The evidence for climatic change in northern Cumbria is derived from twenty-one peat profiles which were extracted from several

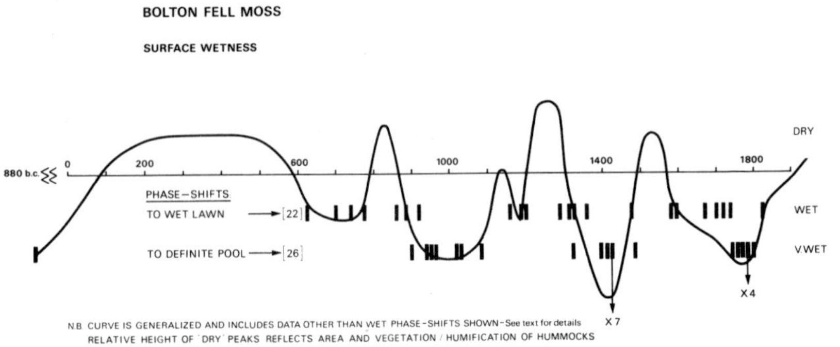

Figure 1 - Surface wetness curve from Bolton Fell Moss, Cumbria, based on macro-fossil analysis of the upper peat and showing a close correspondance with known climatic change.

representative peat faces and subjected to macrofossil analysis. Due to the excellent state of preservation of the peat at Bolton Fell Moss these macrofossils - mainly leaves of *Sphagnum* bog-mosses, generally 1-2mm in size, twigs and leaf-bases of heather and cotton-grass accurately reflect the life-assemblages of the bog vegetation in a continuous sequence over several centuries. Since each species, especially of *Sphagnum*, has a known micro-habitat on the bog related primarily to water-levels, each detailed macrofossil diagram (see Barber, 1980, Figs. 48-68) can be interpreted in terms of the surface wetness of the profile at the time of peat formation. As the bog is ombrotrophic such surface wetness changes must be overwhelmingly due to climatic changes - for various reasons few, if any, changes can be ascribed to changes in hydrology within the bog.

It will be obvious that the only changes which can be accurately dated, through radiocarbon dating and correlated pollen diagrams, are changes to a wetter phase of bog growth. These take place rapidly, within a decade or two, whereas the changes to a drier phase of bog growth can take several decades and can be due to a drier and/or warmer climate *or* to peat building-up above the water-table in a constantly wet and/or cool climate. The basis of the reconstructed surface wetness curve shown as Figure 1 is therefore the "phase-shifts" to either a wet *Sphagnum* lawn type of vegetation or to definite pools of open water. Both types have their characteristic macrofossil assemblages and can be easily discerned in the field stratigraphy.

Figure 1 is based on a summation of the changes recorded in 21 peat profiles up to 1.50m deep in which there occurred 48 phase-shifts. The dates and type of shift are marked along a time-axis stretching continuously from 0-1950 A.D. with a single outlying shift at 880 B.C. This early shift may equate with the well-known climatic change marked by the *Grenzhorizont* in many northwest European bogs. Thereafter the

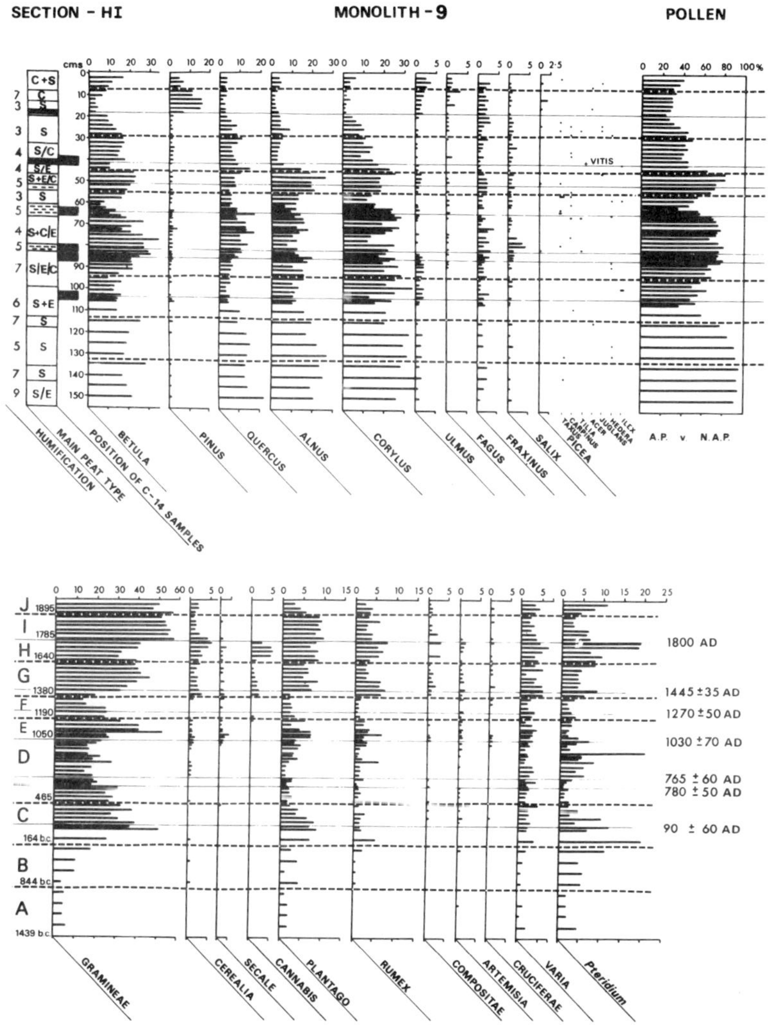

Figure 2 - The main pollen diagram from Bolton Fell Moss, Cumbria - a continuous record of land-use change since pre-Roman times, based on pollen sum of at least 250 non-mire pollen. A summary diagram showing main types only.

peat at Bolton Fell Moss grew up in a dry phase with much heather until shortly after 600 A.D. The relative heights of the "dry" phases on the curve (though the bog would still seem relatively wet to a layman) are based on considerations of the humification of the peat and its macrofossil content as well as the absence of "wet" indicators.

After 600 A.D. the curve gives a continuous record of climatic fluctuation which agrees strikingly with the evidence summarized by Lamb (1977). We can see that the minor deterioration from 600-800 A.D. reversed for a few decades in the 800s, is followed by about two centuries of cool and/or wet climate superseded by the well-known Medieval Optimum. This is brought to an end by a catastrophic decline, beginning in the early 1300s and reaching a nadir in the early 1400s - seven shifts to open pools are dated at 1425 A.D. This is, of course, in accord with much evidence for the opening of the Little Ice Age, including that of Parry (1975), also from the Anglo-Scottish Border area. The amelioration during the early 1500s, a phase of minor hummock-building and no pools on Bolton Fell Moss, has been noted by several authors (see, for example, Lamb, 1977, p 461), and is followed by a return to wetter peat formation culminating in widespread pool formation during the later 1700s - four pools date from 1780 A.D.

The pollen evidence (Fig. 2) cannot be treated in this way. What we have here is a continuous record of agricultural change from pre-Roman Times to the present day and whereas pollen changes have been used to deduce climatic changes during the earlier postglacial (see especially Frenzel, 1966), the factors at work on the dry-land vegetation around the bog during the last two millenia are altogether more complex than the simple relationship which existed on the bog itself. There is in fact a close correspondence between recorded and surmised land-use change in northern Cumbria and the pollen diagram reproduced as Figure 2. The Roman impact shows up a marked decline in trees (all on the left of the diagram, and in the A.P. - arboreal pollen - against N.A.P. - non-arboreal pollen - curve) and an increase in grasses (*Gramineae*), plantain (*Plantago*) and bracken (*Pteridium*). The site is only some 5 km north of Hadrian's Wall. Then there is a period of declining agricultural activity until the Norse settlement (late zone D and zone E) followed by a sharp decline in the period 1190-1380 A.D. which must be mainly attributable to the Anglo-Scottish border troubles and the Black Death. Thereafter the picture is one of almost unremitting agricultural advance except for a short period around 1640, possibly connected with the Civil War.

It can be seen therefore that any attempt to invoke climatic causes in an interpretation of this pollen diagram could be dangerously misleading. One may surmise that the slightly better showing of elm pollen (*Ulmus*) between 100 and 700 A.D., and again in the 1200s could be a consequence of a drier and/or warmer climate but one cannot extend this argument to the top of the diagram when along with pine (*Pinus*) and beech (*Fagus*) the tree was being planted on country estates. Similarly the single grain of the vine (*Vitis*) recorded at about 1450 A.D. cannot be used as a climatic indicator - it occurs during the period of maximum surface wetness on the bog and, as the late Professor Gordon Manley remarked to me: "...a lot of monks and a lot of bricks ....''. Lanercost Priory is only 7 km to the southwest. One might also note that the greatest decline in alder (*Alnus*), a tree of notably wet soils, occurs during the early 1400s - the reverse of what one would expect on climatic grounds but in accord with the increasing tempo of agricultural change.

It is therefore clear that the dual record of peat and pollen stratigraphy of the uppermost horizons of raised bogs can be of value in tracing the climatic and agricultural record of the last few millenia. This is by no means a new idea - the publications by Turner (1970) and by Oldfield (1969), on closely-counted pollen diagrams, extending up to the present bog surface, summarize their earlier pioneering work - and Turner's more recent work (see, for example, Turner, 1979) will be of interest to historical geographers concerned with the north of England. The author is presently engaged on the same approach in southern England, particularly in the New Forest (Barber, 1975). On the peat stratigraphical and macrofossil side there seems, by contrast, to be little work in progress, and the sites are disappearing fast due to exploitation of the upper peat for *Sphagnum* moss garden litter. Since the record from Bolton Fell Moss shows such good correspondence with what is known from England and Wales as a whole it may be that further work on the peat/climate record would only duplicate the results of Figure 1, and in any case it is difficult to think of suitable sites in lowland England and Wales. The potential of lowland and highland Scotland for such work is, however, very high - a suitable site in the Lammermuir Hills would tie up nicely with Parry's work (1975).

It has been shown here that the consequences of climatic deterioration in the Cumberland lowland were to change the type of raised bog vegetation on a large scale. Ground water tables must also have been raised and growing conditions must have been adversely affected to some degree but this certainly does not show up in the Bolton Fell Moss pollen diagram - agricultural indicators in the diagram peak at times when the bog was wettest, around 1050 A.D. the 1400s and around 1800 A.D. The consequences of climatic change were therefore ecological changes on a sensitive vegetation type, changes not apparently of a magnitude to affect agriculture adversely - or at least not more so than the Scots. It would be interesting to see the results of a study such as this on sites further north - and higher up.

REFERENCES

Barber, K. E. (1975) Vegetational History of the New Forest: a preliminary note. *Proceedings of the Hampshire Field Club and Archaeological Society*, 30, 5-8.

Barber, K. E. (1976) History of Vegetation. In S B Chapman (ed.) *Methods in*

Plant Ecology, 5-83, Blackwell Scientific Publications, Oxford.

Barber, K. E. (1978) *A palaeoecological test of the theory of cyclic peat bog regeneration*. Ph.D. thesis, University of Southampton (2 vols).

Barber, K. E. (1981) *Peat Stratigraphy and Climatic Change: a palaeo-ecological test of the theory of cyclic peat bog regeneration*. Balkema, Rotterdam.

Frenzel, B. (1966) Climatic change in the Atlantic / sub-Boreal transition on the northern Hemisphere: botanical evidence. In *World Climate from 8000-0 BC*. Royal Meteorological Society, London.

Lamb, H. H. (1977) *Climate: Present, Past and Future*. Vol. 2: *Climatic history and the future*. Methuen, London.

Oldfield, F. (1969) Pollen analysis and the history of land-use. *Advancement of Science*, London 25, 298-311.

Parry, M. L. (1975) Secular climatic change and marginal agriculture. *Transactions, Institute of British Geographers* No 64, 1-13.

Turner, J. (1970) Post-Neolithic disturbance of British vegetation. In D Walker & R G West (eds.), *Studies in the Vegetational History of the British Isles*, 97-116. University Press, Cambridge.

Turner, J. (1979) The Environment of northeast England during Roman Times as shown by Pollen Analysis. *Journal of Archaeological Science* 6, 285-290.

MARGINAL LAND AND CLIMATE; A COMMENT

Adrian H. Cowell
Aughton Green, Lancashire

The circumstances in which upland settlements are established and deserted are complex. There would appear to be evidence from upland Wales to suggest that fundamental changes in the social structure of pastoralism had subtle long term effects upon their ecology, economic viability, and long term fertility.

There would appear to be three major explanations for marginal land. As well as environmental explanations there are also economic and geographic explanations and these account for most of the 'marginal' land in Great Britain. Perhaps only at the highest altitudes of our mountain ranges would climatic explanations alone explain changes in settlement or desertion. In Scotland, at least, this would appear to be the case if soils are regarded as a reliable guide for Birse (1980) has suggested that the World Soils Classification should be modified to allow for soils in the "Oro-arctic" where rapid temperature fluctuations seem to explain one type of soil whilst wind blown aggregates suggest the further category of Aeolisols. To a great extent such soils originate from the adverse climatic regime which hinder organic degradation and equally retard plant growth. The accounts of such areas in Iceland and Norway by Ogilvie and Grove (this volume) indicate the narrow range of tolerance for human settlement may be entirely attributed to climatic vagaries. This does not seem to be the case in the areas studied by Beresford or White (this volume) where other factors are arguably more significant in explaining the expansion or contraction of agricultural endeavour and human settlement. The comments on Beresford's paper by Dr. Delano Smith concerning a possible confusion between desertion of habitable dwellings and desertion of agricultural terrain are of material importance in this respect.

In the sub-montane grasslands (Evans, Hill and Ward, 1977) where grazing is predominant but not all pervasive there would appear to be episodes when arable farming has occurred on a wider scale than can be explained by the provision of subsistence needs or winter forage and some of these episodes may be attributed to climate as is suggested by the work of Moore and Chater (1969) and by Turner (1964). The need for caution in these matters has been clearly discussed by Parry in his study of methodology (this volume) and his pertinent view that climatic episodes might be properly seen as 'trigger mechanisms' which either compound or exacerbate problems created by other factors is especially important in the areas studied by White and Beresford. Of special importance in the understanding of the sub-montane grassland areas is the problem of fertility - the agricultural equivalent of 'potential energy'. This potential energy is easily dissipated unless nutrients are refurbished and time is allowed for natural recuperation. If either of these precautions are not observed the land has limited life as a resource for permanent settlement base on arable or livestock enterprise. It may be postulated that it is precisely in those areas which are marginal for economic or geographic reasons that climate operates according to Parry's scheme rather than as a unique harbinger of disaster. Fundamental changes in social structure or in legal or political institutions can have seminal effect although less directly and perhaps over rather longer time scales than climate.

It is a matrix of social and ecological factors which must be invoked to explain the "desertion" of land in the sub-montane grasslands of Celtic Britain between the sixteenth and twentieth centuries. In these areas, which were on the periphery of a newly burgeoning economy, societies with quite distinct social institutions practised a quite distinct form of agriculture and did so often in areas where the geology had provided a poor quality of soil which was essentially acid and poor in exchangeable bases. High rainfall and drainage could indeed contribute towards the problem by leaching nutrients out of the fertile horizons into the less accessible levels and eventually into the streams and rivers. The implications of rainfall for soils and vegetation and the effect of these on human settlement has been studied by Hughes (1946). Individual soils have their own critical point at which their fertility is impaired by a combination of factors including rainfall and drainage but this critical point may be only precisely identified in situ. The importance of such nutrients for the sward has been elucidated by Milton (1940) who has indicated how applications of P.K.N. and Ca. can restore an impoverished upland sward of *Nardus stricta* and *Molinia caerulea* to a condition comparable with a good lowland ley in which lowland species were quite quickly recruited into the plant community. Studies in sheep behaviour revealed also the special selectivity of sheep grazing and indicates the circumstances in which particular species may be acceptable or not: equally the variety of sward "on offer" was shown to be important and was proven to be a considerable feature in grazing management. The interaction between the sward and the grazer is of critical importance as the work of Boulet (1940) has shown. The evidence of a mosaic of over- and under-grazed swards indicates intensive stocking of sheep as well as the neglect of cattle. Roberts (1959) has suggested that enclosure during the eighteenth century was primarily responsible for the deterioration of Welsh sub-montane grasslands during the nineteenth century. He equally asserts that the intensive stocking of sheep in Wales was a relatively new phenomenon in Wales and occurred only during the sixteenth century and after. This author believes that the ecological evidence available both from Wales and Scotland suggests an alternative thesis.

It was Boulet's scheme to distinguish the management of grassland into two "complex"; a "sward complex" and a "grazing complex" which may be analysed individually before being placed together in order to elucidate the interactions between them. As in the case of Milton this is possible if grazing is controlled or restricted completely, and it is clear from Milton's work that recruitment of lowland species into an improved sward occurred more rapidly if the grazing regime was totally abolished. More realistically he was able to show that controlled grazing achieved similar results and a fortunate coincidence allowed him also to recognise the significance of cattle in a grazing scheme. The mixed grazing stock of cattle and sheep in an almost tribal transhumance has been regarded as an established feature of Welsh grasslands until the decline of the Gafael as a result of the Statute of Ludlow and the Welsh Act of Union. Jones-Pierce (1942) has been able to demonstrate this decline of Gafael in Caernarvonshire in the late fifteenth century and other documents demonstrate the same process accelerating during the sixteenth and seventeenth century. It may be argued that such a pattern of transhumance in which the land is held in common by an extended group, family, or tribe would allow grazing to occur over a wide variety of sward standard. During the summer temporary higher pasture (Hafod) would be grazed between May and October and this allowed the required time for the grass to recover whilst mixed stock maintained over a period of years averted the worst effects of nutrient impoverishment. The enclosures of the sixteenth century were associated with a process of encroachment upon common grazing both by those legally entitled so to do and those who were (as in White's example) illegally squatting. These encroachments, usually into the best quality land on the former "Hafod", resulted in the splitting up of the various qualities of sward as well as the permanent settlement of the land. Such a process must be regarded as ecological folly and could be expected to offer little prospect of long term success. A series of poor years through climatic inclemency might indeed precipitate a small farmer into an irreversible, and irresistable, spiral of decline but so could any financial crisis as a consequence of such episodes as the endemic of Rinda Pest (Cattle Fever) in the eighteenth century, or a series of uncontrolled events along the drover routes. Rees (1929) and Price-Edwards (1932) have mentioned the strange effects upon agriculture as a result of the railway penetrating into Wales both in its direct effects and its indirect effects upon the coastal shipping traffic during the nineteenth century. The period of great rural discontent known as the Rebecca Riots has been analysed by Williams (1955) and he can point to the consequences of the turnpike trusts upon the importation of lime and potash, which made the movement of such dressings prohibitive.

The alternative thesis to that suggested by Roberts, already referred to, suggests that in the marginal areas the impact of an expanding and competitive economy upon an essentially subsistence agriculture was disastrous. Where a tribal system such as that implicit in a clan or gafael could provide mutual support and buffer the worst effects of financial difficulties, the small proprietors which replaced the primitive co-operatives had neither the financial means nor the economic or ecological resilience to resist the effects of the more vigorous economic neighbour. What is now being suggested is that in addition to the two complexes identified by Boulet there should be envisaged a third "Social complex" which has an important ecological significance because of the collective structure of the clan in contradistinction to the individualistic private land owner who followed in the wake of the new Tudor dispensation in Wales (Elton 1967). Such a collective structure could provide a necessary ecological buffer because the society was organised in such a way as to absorb the worst effects of short term exigencies as well as providing a long term precaution against dissipating fertility by ensuring the presence of mixed grazing to meet the immediate needs of the society. This would have the advantage of countering the worst effects of external and largely uncontrolled market demands.

To recapitulate; there are many ecological perturbations which may have been responsible for the changes in agricultural practice and the corresponding changes in settlement patterns. In the sub-montane grasslands of Britain between 1600 and 1940 many of the economic changes and their social consequences significantly modified deleteriously the ecology of the managed system so that climatic change may be seen to have either accelerated or delayed the inevitable processes of decline initiated by severing the link between social structure and ecological equilibrium. The extent to which this may be seen is demonstrated succinctly by Mutch in his study of Fassfern in Aberdeenshire. The caution advocated by Parry in the association of climate with agricultural desertion would seem to be strongly supported by ecological evidence.

REFERENCES

Birse E. L. (1980) Suggested amendments to the World Soil Classification to accommodate Scottish Mountain and Aeolian soils. *Journal of Soil Science* 31, 1, 117-24.

Boulet, J. L. (1940) *The Ecology of a Sheep Walk in Mid-Wales.* Ph.D. Thesis (Unpublished), University College of Wales, Aberystwyth.

Elton, G. R. (1967) *Tudor Constitutional Documents.* Cambridge University Press.

Evans, D. E., Hill, M. O. and Ward, S. D. *A Dichotomous Key to British Sub-montane Vegetation.* Institute of Terrestrial Ecology, Bangor, Occasional Paper No. 1.

Hughes, R. E. (1946) *Plant Ecological Studies in the North West Conway Valley, with reference to a parallel study of human settlement.* Unpublished Ph.D. thesis, University College of North Wales, Bangor.

Jones-Pierce, T. (1942) The Gafael in Bangor Manuscript No. 1939 circa 1400. *Transactions of the Honourable Society of Cymmrodorion* 158-88.

Milton, W. E. J. (1940) Grazing and species composition. *Journal of Ecology*, 38, 326-356.

Moore, P. D. and Chater E. W. The changing vegetation of west central Wales in the light of human history. *Journal of Ecology*, 57, 361-79.

Mutch, W. E. S. (1972) *Study of combined land uses. Fass Fern, Invernesshire.* University Press, Edinburgh.
Price-Evans, E. (1932) Cader Idris: a study of certain plant associations in south west Caernarvonshire. *Journal of Ecology*, 20, 1-52.
Rees, J. (1929) *The Rough Pastures of the Llyn Peninsula*. Unpublished M.Sc. Thesis, University College of North Wales, Bangor.
Roberts, R. A. (1959) The ecology of human occupation and land use in Snowdonia. *Journal of Ecology*, 47, 2, 316-323.
Turner, J. (1964) The anthropogenic factor in vegetation. *New Phytologist*, 63, 73-90.
Williams, D. (1955) *The Rebecca Riots*. University of Wales Press, Cardiff.

COMMENTS

Bruce Proudfoot,
Department of Geography,
University of St. Andrews

At the Conference, Professor Proudfoot was invited to conclude the proceedings. The gist of his summary is contained in the following Comment (eds.).

To implicate climate in the events of the past, and especially to evaluate its role in changes in agriculture and rural settlement, requires that the relationships between climate and agricultural production (for example, crop yields) be established. It must be stressed that relationships between present day crop yields and climate have proved difficult to establish. Sophisticated experiments in phytotrons, which clearly demonstrate relationships between yields and environmental conditions, cannot be replicated under field conditions, for the real weather during any growing season is notoriously fickle, the representativeness of meteorological station data for wider field conditions is often uncertain, and variability in soils and in farm management introduce further complexities. Comparisons of crop yield data and climatic data over many years in any sizeable region are difficult since crop locations, acreages, and varieties change, as does the technology of crop production. (Climatic variation over time may be quite different in the various smaller areas that go to make up the larger region, as Gregory clearly illustrated many years ago in his series of maps of annual variations in the climate of northwestern Europe (Gregory, 1954). Data on both climate and crop yields collected over a number of years are likely to represent short term variations as well as longer term trends; statistical methods for the handling of such mixed data are still a matter of debate amongst statisticians.

Climate and weather are such obvious factors in crop production that they are often assumed to be the most important. However, as is made clear in several of the contributions to the present volume, the many non-climatic conditions sufficient to account for agricultural change have to be constantly borne in mind. For instance, it is arguable that at least for some types of crop production the Climate Problem in Upland Britain is as much a fertility problem as a climate problem. Leaching of plant nutrients, and the harvesting of crops and livestock from many upland areas over millennia have so seriously depleted available plant nutrients that sustained yields can now be obtained only by the regular addition of fertilisers. Disruption of an agricultural system that depended on applications of manure to cultivated fields in the uplands might very rapidly lead to the abandonment of cultivation without any significant change in climate having taken place. The re-allocation of resources could make lowland farming more attractive than that at higher altitudes, as Whyte suggests for parts of southeast Scotland.

In the field, and indeed in much documentary evidence, it is changes in settlement or crop yields along the limits of the settled area that can be most clearly demonstrated, as we have seen in the papers on Scotland, Iceland and Norway, where, intuitively, we expect that climatic change would have most impact. It is clear, however, that many such peripheral areas are marginal not only in a physical sense but also in an economic sense. Quite small changes in economic circum-stances may be just as disastrous as any changes in climate. However, bad weather which reduces yields may also lead to higher prices for the producer so whether or not a commercial producer under such circum-stances is better or worse off will depend on the volume of crop produced and the price received per unit volume. There is, in short, no single, simple, necessary outcome to this sort of situation. Persistence of unfavourable conditions over a number of years will affect the farmer with adequate resources (often financial, but sometimes also off-farm resources) very differently from the near-subsistence farmer producing just enough to survive for a single season. The contribution of peripheral farmers to the total economy may be so small as to be economically insignificant, although their plight may be so newsworthy that their situation is heavily over-represented in the documents of the time.

It might be argued that such comments are irrelevant to earlier periods of European agriculture when agriculture was essentially subsistence in nature. This is surely an out-moded simplistic view. Goods were priced and traded and taxes collected in kind or cash from early medieval times, as documentary and, increasingly, archaeological evidence indicates. West European agriculture in the historic period cannot be regarded as an isolated element of the economy responding in a simple direct manner to the changes of climate, and the variations in weather from year to year that certainly occurred. Population change, migration, the development of towns, changing social and political institutions were factors influencing, and being influenced by, agriculture. Something of the complexity of these interrelationships on a broad European scale can now be more widely appreciated with the recent publication in English of Abel's *Agricultural Fluctuations in Europe* which complements Slicher van Bath's earlier *Agrarian History of Western Europe*. These should make us critical of simplistic explan-ations of changes in house forms, building styles or settlement distributions solely in terms of climatic change. This is not to deny the reality or signficance of climatic change, rather to stress the complexities of individual situations.

It is also important to note, from the climatological point of view, the divergence of opinion on the nature of climatic change and its causes. Some have argued that the cyclic nature of climatic variations can be linked to a similar variation in sunspot cycles, while others agree that such cyclic patterns can be found even though they find little convincing evidence that the cycles are related to each other in any casual, or even statistical, sense (see, for example, Proceedings Symposium/Workshop, *Solar-terrestrial influences on weather and climate*, Columbus, Ohio, 1978, eds. B. M. McCormac and T. A. Seliga (Reidel) 1979). Dury's argument that step-functional change of climate is normal provides a more structured format for the frequently argued proposition that runs of good years tend to be succeeded by runs of bad years. The extent to which short-term changes of this sort are superimposed on, or contribute to, long-term change is yet another unresolved problem.

The rapidity and reality of climatic change are clearly demonstrated during the last two centuries in north American context. In this respect Lawson's paper usefully demonstrates the way in which biological, tree ring, evidence for climatic change could aid in understanding the historical records. The apparent conflicts in this record can partly at least be explained by the rapid changes in landscape produced by climatic change. Changing climatic conditions in the early nineteenth century might well have had an exceptional impact upon existing plant systems because of instability produced during the previous two centuries. Already by 1820, when Long named the Great American Desert, the traditional economic and environmental systems had been subject to change. The arming of the native population and the acquisition of the horse had upset traditional balances between neighbouring population groups, and between populations and plant and animal resources. Under such changing and unstable environmental circumstances drier than usual years may have had a greater than usual effect. In a similar way the effects of the dry years of the late 1920's and 1930's were intensified in the western grain-growing plains of the United States because of the narrow range of crops that had succeeded a floristically richer and more diverse system. In the area to the north of the Great American Desert, in the southern Prairies of Canada, where the same conflicts of historical evidence as to environmental conditions are found, the dry years of the 1920's and 1930's were not the sole cause of rural distress. As significant were the low prices obtained by the farmer when he took his meagre crop to the market for sale. Significantly, during the last half century the development of new agricultural techniques has enabled prairie grain farmers to produce crops during years as dry as those of the 'Dirty Thirties'. True, some of the driest areas were abandoned for tillage in the 1930's; but, equally significant, many farmers from these areas moved north to the then frontier of settlement and found, within a distance of no more than 300 miles from their dried-out farms, areas where rainfall was above average at that time. Whereas grain prices were everywhere low the incidence of drought varied regionally.

Reductions in crop output resulting from unfavourable weather conditions may occur on a large scale, as is well seen in Canadian Wheat Output which has shown reductions of as much as fifty per cent between successive years. In 1942 output was 556M. bushels; in 1943, 284M. bushels; in 1952, 688M. bushels; in 1954, 331M. bushels. Swings of such magnitude within a few years seem in recent times to have had little impact on the core production areas, but to have emphasised rather the fragility (in both physical and economic terms) of peripheral marginal producers. For the longer historical perspectives adopted in these papers, could the same conclusions be applied? Are the effects of climatic change and of economic fluctuations, whether acting singly or together, greatest at the limits of any productive activity? To what extent is the evidence from marginal locations likely to provide exaggerated evidence for change, whether or not such change may be attributed to climate. To what extent can the factors of change recognised in marginal areas, precisely because the evidence is exaggerated, be applied to all areas?

REFERENCE

Gregory, S. (1954) Climatic Classification and Climatic Change. *Erdkunde*, 8, 246-252.

Text Processing by

Security Word Processing
5 Cheslyn Drive, Aspley Hall, Nottingham
England

Printing by

Department of Geography
The University of Nottingham
Nottingham, England

Papers arising
from a meeting of the
Historical Geography Research Group,
Institute of British Geographers,
at the University of Nottingham
11th to 13th July 1980